Universitext

Universitext is a series of textbooks that presents material from a wide variety of mathematical disciplines at master's level and beyond. The books, often well class-tested by their author, may have an informal, personal even experimental approach to their subject matter. Some of the most successful and established books in the series have evolved through several editions, always following the evolution of teaching curricula, into very polished texts.

Thus as research topics trickle down into graduate-level teaching, first textbooks written for new, cutting-edge courses may make their way into *Universitext*.

Rinat Kashaev

A Course on Hopf Algebras

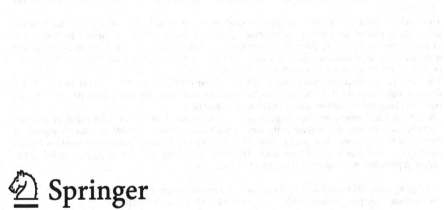 Springer

Rinat Kashaev 🆔
Section de mathématiques
University of Geneva
Genève, Switzerland

This work was supported by Schweizerischer Nationalfonds zur Förderung der Wissenschaftlichen Forschung (200020_192081, 200020_200400), National Centres of Competence in Research SwissMAP (NCCR The Mathematics of Physics)

ISSN 0172-5939   ISSN 2191-6675   (electronic)
Universitext
ISBN 978-3-031-26305-7   ISBN 978-3-031-26306-4   (eBook)
https://doi.org/10.1007/978-3-031-26306-4

Mathematics Subject Classification: 16Txx, 15-XX, 18Mxx, 57Kxx

This Springer imprint is published by the registered company Springer Nature Switzerland AG
The registered company address is: Gewerbestrasse 11, 6330 Cham, Switzerland

To the memory of my parents,
Mavlavi Kashaev and Lena Urazaeva

Preface

The book is based on the one-semester introductory courses on Hopf algebras that the author taught on several occasions at the mathematical department of University of Geneva, and it is addressed mostly to those who learn the subject for the first time. Compared to the already existing vast literature on the subject, the distinguishing features of this book are as follows:

- We restrict ourselves to only minimal amount of material needed for applications in knot theory, namely, construction of solutions of the Yang–Baxter equations.
- The presentation is purely algebraic, essentially based on the multilinear algebra.
- When possible, the string diagram notation for monoidal categories is used which facilitates many proofs.
- The restricted (or finite) dual of a Hopf algebra plays the central role.
- The general theory is developed in a minimal way, while few simple examples are considered and worked out in detail.
- The universal R-matrix is given in dual terms as a bilinear form, so that for infinite dimensional Hopf algebras no completion is needed.

The main motivation that guides this book is the construction of solutions of the Yang–Baxter equations which are written graphically as follows:

$$\sum_{p,q,r\in E}\quad\begin{array}{c}j\quad k\\[-2pt]\nwarrow\quad\nearrow\\ q\diagup\backslash r\\ i\leftarrow\diagup\backslash\quad n\\ \diagup\quad p\quad\backslash\\ l\qquad m\end{array}\quad=\quad\sum_{s,t,u\in E}\quad\begin{array}{c}j\qquad k\\ \nwarrow\quad\nearrow\\ i\leftarrow\diagdown s\diagup\quad n\\ u\diagdown\diagup t\\ \diagup\backslash\\ l\quad m\end{array}$$

where the labels, taking their values in a finite set E, are assigned to edges so that the same labels are assigned to the corresponding open ends of two sides of the graphical equations. The unknowns correspond to the crossings with the labels around:

$$\underset{k}{\overset{i}{\diagdown}}\!\!\!\underset{l}{\overset{j}{\diagup}} = x_{i,j,k,l}$$

whereas the contribution of each labeled diagram is given by the product of the unknowns over all crossings. So, explicitly, the equations are as follows:

$$\sum_{p,q,r\in E} x_{j,k,q,r} x_{i,q,l,p} x_{p,r,m,n} = \sum_{s,t,u\in E} x_{i,j,u,s} x_{s,k,t,n} x_{u,t,l,m}, \quad i,j,k,l,m,n \in E.$$

As there are $|E|^6$ nonlinear equations on $|E|^4$ unknowns, this system of equations is overdetermined if $|E| > 1$.

The Yang–Baxter equations were first introduced in the field of two-dimensional lattice statistical mechanics and quantum field theory as an algebraic machinery for treating exactly solvable models [6, 12, 47]. Their name comes from the works of C.N. Yang [46] and R.J. Baxter [5], though the equations first appeared in the work of J.B. McGuire [30], and their specific form, known as star-triangle relations, is already contained in the work of L. Onsager [33].

In the 1980s, a revolutionary breakthrough in three-dimensional topology took place with the discovery by V.R. Jones of a new polynomial invariant of knots and links [18] which appeared to be related with the solution of the Yang–Baxter equations associated to the six-vertex model of lattice statistical mechanics. The discovery of the Jones polynomial has resulted in creation of a new mathematical subject called Quantum Topology where the Yang–Baxter equations play the central role as concerns the construction of knot and link invariants. The subsequent development of the theory of quantum groups [11, 13, 17] has led to deep connections of the Yang–Baxter equations with Hopf algebras, and since then the so-called quantum invariants have proliferated and their study has been closely tied with connections between monoidal categories and 3-manifolds [41] and the development of topological quantum field theories [3, 38, 45].

The outline of the book by chapters is as follows.

In Chap. 1, after brief discussion of the setup and some preparatory work with the definition of a group and the graphical notation of string diagrams, we give the definitions of a Hopf algebra and related algebraic structures: algebras, coalgebras, bialgebras, modules, and comodules, and discuss their general properties.

In Chap. 2, we describe how to construct Hopf algebras based on the constructions of algebras, coalgebras, and bialgebras. Particular attention is paid to free algebras on vector spaces and free bialgebras and Hopf algebras on coalgebras.

In Chap. 3, we introduce the restricted (or finite) dual of an algebra. It is a coalgebra that accumulates all finite-dimensional representations of the algebra. This is one of the most important ingredients for the subsequent chapters.

In Chap. 4, we work out in detail the structure of the restricted dual for few specific Hopf algebras.

In Chap. 5, we define Drinfel'd's quantum double $D(H)$ of a Hopf algebra H in its maximal form by using the restricted dual of H and describe its structure in the case of the quantum group B_q, a Hopf algebra isomorphic to the Borel subalgebra of $U_q(sl_2)$. We also work out in detail the calculation of the solutions of the Yang–Baxter equations associated with irreducible finite-dimensional representations of $D(B_q)$.

The last Chap. 6 is dedicated to construction of knot invariants from solutions of the Yang–Baxter equations. In particular, we define a representation independent universal knot invariant associated to a given Hopf algebra H with invertible antipode. This invariant takes its values in the dual space of the restricted dual of the quantum double of H.

The presentation of the material of Chaps. 1–5 is based mainly on the books [1, 10, 21, 29, 35, 39]. The last Chap. 6 can be complemented with some standard introductory textbook on knot theory, for example, [23].

I would like to thank Remi Lodh and the whole team of Springer publishers/editors and the anonymous referees for their help, advice, and numerous comments which greatly improved the presentation of the book's content. I am grateful to my colleagues and students for valuable discussions related to the contents of this book, especially to Jørgen Ellegaard Andersen, Vladimir Bazhanov, Anna Beliakova, Christian Blanchet, Stavros Garoufalidis, Nicolas Hemelsoet, Mucyo Karemera, Kalle Kytölä, Vladimir Mangazeev, Marcos Mariño, Jules Martel-Tordjman, Cristina Palmer-Anghel, Martin Palmer, Eiichi Piguet-Nakazawa, Jan Pulmann, Muze Ren, Nikolai Reshetikhin, Louis-Hadrien Robert, Sergey Sergeev, Shamil Shakirov, Sakie Suzuki, Vladimir Turaev, and Alexis Virelizier, among others.

This work is supported in part by the Swiss National Science Foundation, the subsidies no 200020192081, 200020200400, and the research program The Mathematics of Physics NCCR (SwissMAP).

Geneva, Switzerland   Rinat Kashaev
December 2022

Notation and Conventions

For any non-negative integer $n \in \mathbb{Z}_{\geq 0}$, the corresponding von Neumann ordinal number, that is the finite set $\{0, 1, \ldots, n-1\}$, is written as underlined n:

$$\underline{n} := \{0, 1, \ldots, n-1\}.$$

For example, $\underline{0} = \emptyset = \{\}$, $\underline{1} = \{0\}$, $\underline{2} = \{0, 1\}$, etc.

We also use ω as an alternative notation for the set of non-negative integers $\mathbb{Z}_{\geq 0}$.

For any set S, the cardinality of S is denoted by $|S|$. For example, $|\underline{n}| = n$ for any $n \in \omega$.

For any two sets X and Y, X^Y denotes the set of all mappings from Y to X. For example, X^ω denotes the set of sequences $(x_0, x_1, \ldots)$ in X, while the set $X^{\underline{n}}$ is naturally bijective to the Cartesian power X^n through the map $X^{\underline{n}} \to X^n$ that sends a function $x \colon \underline{n} \to X$ to its ordered image $(x_0, x_1, \ldots, x_{n-1}) \in X^n$.

If sets $X, Y, \ldots, Z$ are finite or infinite countable, then for any map

$$f \colon X \times Y \times \cdots \times Z \to W,$$

we often write $f_{x,y,\ldots,z}$ instead of $f(x, y, \ldots, z)$.

An unspecified base field is denoted as $\mathbb{F}$, while for specific examples we often work with the field $\mathbb{C}$ of complex numbers or the field $\mathbb{R}$ of real numbers.

For an $\mathbb{F}$-linear map $f \colon V \to W$ between vector spaces over a base field $\mathbb{F}$, we write fx instead of $f(x)$. For an $\mathbb{F}$-linear form $f \colon V \to \mathbb{F}$, we also write $\langle f, x \rangle$ instead of fx.

For a composition of linear maps, we suppress the composition symbol and write fg instead of $f \circ g$.

For two vector spaces U and V (over a field $\mathbb{F}$), the notation $\mathrm{L}(U, V)$ is used to denote the vector space of all $\mathbb{F}$-linear maps from U to V.

Contents

Chapter 1
Groups and Hopf Algebras

The main goal of this chapter is

- to give the definition of a Hopf algebra and to motivate it on the basis of the notion of a group which is of fundamental importance in mathematics;
- to introduce the graphical notation of string diagrams;
- to introduce the algebraic structures closely related to Hopf algebras, namely the notions of algebra, module, coalgebra, comodule, convolution algebra, and bialgebra;
- to establish few basic properties of Hopf algebras.

Section 1.1 in this chapter is the most abstract one where we briefly discuss the notions of a monoidal category, a braided monoidal category and a symmetric monoidal category. We do so for at least three reasons. First, those notions are used in the last Chap. 6 where knot invariants are defined in the general context of monoidal categories. The second reason is that, we motivate the definition of a Hopf algebra by the definition of a group (reformulated by using the structural maps), and these two definitions differ only by the underlying symmetric monoidal categories: vector spaces with tensor product in the case of Hopf algebras and sets with Cartesian product in the case of groups. The third reason is that many general constructions and statements in the book can be expressed in the language of string diagrams, and the latter make sense also in the context of arbitrary symmetric monoidal categories. In principle, with the exception of the last Sect. 1.1.3, where the graphical notation of string diagrams is introduced, Sect. 1.1 is optional for five Chaps. 1–5 where we mainly work only in the framework of multilinear algebra, that is the symmetric monoidal category $\mathbf{Vect}_{\mathbb{F}}$ of vector spaces over a field $\mathbb{F}$ with the tensor product $\otimes_{\mathbb{F}}$ as the monoidal product. Thus, one can start the reading right from Sect. 1.1.3 and return back to Sect. 1.1 only before reading Chap. 6 dedicated to applications in knot theory.

© The Author(s), under exclusive license to Springer Nature Switzerland AG 2023
R. Kashaev, *A Course on Hopf Algebras*, Universitext,
https://doi.org/10.1007/978-3-031-26306-4_1

1.1 Monoidal Categories

In this section we give few basic definitions that concern monoidal or tensor categories, without elaborating details but providing few concrete examples. It is assumed that the reader is familiar with definitions of a category, a functor, a natural transformation, and a commutative diagram.

Since our main example of a monoidal category is the category $\mathbf{Vect}_{\mathbb{F}}$ of vector spaces over a fixed base field $\mathbb{F}$ equipped with the tensor product $\otimes_{\mathbb{F}}$ as a monoidal product, on first reading, this section can be viewed as a summary of general properties of the category $\mathbf{Vect}_{\mathbb{F}}$. A more systematic and detailed presentation of monoidal categories can be found in Chapter 1 of the book [42] and chapter XI of the book [19].

1.1.1 Monoidal Categories

For any category $\mathcal{C}$, let $\mathcal{C} \times \mathcal{C}$ be the cartesian square of $\mathcal{C}$, which is a category whose objects are ordered pairs of objects of $\mathcal{C}$, morphisms are ordered pairs of morphisms of $\mathcal{C}$, and the composition is component-wise composition in $\mathcal{C}$.

A category $\mathcal{C}$ is called *monoidal* if it is equipped with a functor

$$\otimes \colon \mathcal{C} \times \mathcal{C} \to \mathcal{C}, \quad (A, B) \mapsto A \otimes B, \quad (f, g) \mapsto f \otimes g, \tag{1.1}$$

called the *tensor* or *monoidal product*, an object I called the *unit* or *identity object*, and natural isomorphisms

$$\alpha_{A,B,C} \colon (A \otimes B) \otimes C \to A \otimes (B \otimes C), \quad A, B, C \in \mathrm{Ob}\, \mathcal{C}, \tag{1.2}$$

$$\lambda_A \colon I \otimes A \to A, \quad \rho_A \colon A \otimes I \to A, \quad A \in \mathrm{Ob}\, \mathcal{C}, \tag{1.3}$$

respectively called *associator*, *left unitor* and *right unitor*, which satisfy two families of *coherence conditions* corresponding to commutative diagrams

$$
\begin{array}{ccc}
 & ((A \otimes B) \otimes C) \otimes D & \\
{\scriptstyle \alpha_{A,B,C} \otimes \mathrm{id}_D} \swarrow & & \searrow {\scriptstyle \alpha_{A \otimes B, C, D}} \\
(A \otimes (B \otimes C)) \otimes D & & (A \otimes B) \otimes (C \otimes D) \\
{\scriptstyle \alpha_{A, B \otimes C, D}} \downarrow & & \downarrow {\scriptstyle \alpha_{A, B, C \otimes D}} \\
A \otimes ((B \otimes C) \otimes D) & \xrightarrow{\ \mathrm{id}_A \otimes \alpha_{B,C,D}\ } & A \otimes (B \otimes (C \otimes D))
\end{array}
\tag{1.4}
$$

and

$$(A \otimes I) \otimes B \xrightarrow{\quad \alpha_{A,I,B} \quad} A \otimes (I \otimes B)$$

$$\rho_A \otimes \mathrm{id}_B \searrow \qquad \swarrow \mathrm{id}_A \otimes \lambda_B$$

$$A \otimes B \tag{1.5}$$

respectively called the *pentagon* and the *triangle diagrams*, and the equality

$$\lambda_I = \rho_I : I \otimes I \to I. \tag{1.6}$$

A monoidal category is called *strict* if the natural isomorphisms α, λ, ρ are identities. It is known that any monoidal category is equivalent to a strict monoidal category, see for the proof, for example, [19].

Example 1.1 The category **Set** of sets is a monoidal category with the Cartesian product as the monoidal product and any one-element set, say $\underline{1} = \{0\}$, as the unit object. This is a prototypical example of a monoidal category. □

Example 1.2 The category **Vect**$_\mathbb{F}$ of vector spaces over a base field $\mathbb{F}$ with $\mathbb{F}$-linear maps as morphisms is a monoidal category with the tensor product $\otimes_\mathbb{F}$ as the monoidal product and the base field $\mathbb{F}$, viewed as a vector space of dimension one, as the unit object. This is the principal monoidal category we will be working with in this book. □

Just for the sake of clarity, to make the abstract definition less abstract, below we give some more less intuitive examples of monoidal categories, although they will not be used in any way in the following.

Before giving the next example, let us recall the definition of the direct sum of a family of vector spaces.

Definition 1.1 Let $\{V_i\}_{i \in I}$ be a family of vector spaces over a fixed base field. The *direct sum* of this family is the vector space $V := \bigoplus_{i \in I} V_i$ of all maps from the index set I to the set-theoretical union of all the vector spaces in the family,

$$x : I \to \cup_{i \in I} V_i, \quad i \mapsto x_i, \tag{1.7}$$

that satisfy the condition $x_i \in V_i$ for all $i \in I$, and x_i is the zero vector of V_i for all but finitely many i's.

For each index $i \in I$, there are two canonical linear maps associated to the direct sum $V = \bigoplus_{i \in I} V_i$ of a family vector spaces indexed by I. These are the projection map

$$p_i : V \to V_i, \quad x \mapsto x_i, \tag{1.8}$$

and the inclusion map

$$q_i : V_i \to V, \quad v \mapsto x, \quad x_j = \begin{cases} v & \text{if } j = i; \\ 0 & \text{otherwise.} \end{cases} \tag{1.9}$$

The projection and inclusion maps satisfy the following relations:

$$p_i q_j = \begin{cases} \mathrm{id}_{V_i} & \text{if } i = j; \\ 0 & \text{otherwise,} \end{cases} \qquad \sum_{i \in I} q_i p_i = \mathrm{id}_V, \tag{1.10}$$

where the sum always truncates to a finite sum when applied to an element of V.

Example 1.3 The monoidal category $\mathbf{Vect}_{\mathbb{F}}^{\mathbb{Z}}$ of $\mathbb{Z}$-graded $\mathbb{F}$-vector spaces is a subcategory of $\mathbf{Vect}_{\mathbb{F}}$ defined as follows.

An object V of $\mathbf{Vect}_{\mathbb{F}}^{\mathbb{Z}}$ is the direct sum of a $\mathbb{Z}$-indexed family of $\mathbb{F}$-vector spaces $\{V_n\}_{n \in \mathbb{Z}}$, while a morphism $f : V \to W$ is a linear map such that $f(V_n) \subset W_n$ for any $n \in \mathbb{Z}$. The tensor product $V \otimes_{\mathbb{F}} W$ of two $\mathbb{Z}$-graded vector spaces is the direct sum of the family

$$(V \otimes_{\mathbb{F}} W)_n = \bigoplus_{k \in \mathbb{Z}} (V_k \otimes_{\mathbb{F}} W_{n-k}), \quad n \in \mathbb{Z}. \tag{1.11}$$

The unit object I is the direct sum of the family

$$I_n = \begin{cases} \mathbb{F} & \text{if } n = 0; \\ 0 & \text{otherwise.} \end{cases} \tag{1.12}$$

The tensor product $f \otimes_{\mathbb{F}} g$ of two morphisms $f : X \to U$ and $g : Y \to V$ is the usual tensor product of linear maps.   □

Example 1.4 Define a category $\mathbf{Mat}(\mathbb{F})$ of matrices over a field $\mathbb{F}$ where the objects are elements of the set of non-negative integers $\omega = \mathbb{Z}_{\geq 0}$ and a morphism $f : m \to n$ is a (set-theoretical) map $f : \underline{m} \times \underline{n} \to \mathbb{F}$. The composition of $f : l \to m$ and $g : m \to n$ is the morphism $g \circ f : l \to n$ defined by

$$(g \circ f)_{i,j} = \sum_{k \in \underline{m}} f_{i,k} g_{k,j}, \quad \forall (i, j) \in \underline{l} \times \underline{n}.$$

This is a strict monoidal category where $m \otimes n = m + n$, the unit object being 0. As $\underline{0} = \emptyset$, for any object $n \in \omega$, there is only one morphism from 0 to n and from n to 0.

For two morphisms $f: m \to n$ and $g: k \to l$, their tensor (monoidal) product is the morphism $f \otimes g: m + k \to n + l$ defined by

$$(f \otimes g)_{i,j} = \begin{cases} f_{i,j} & \text{if } (i, j) \in \underline{m} \times \underline{n}; \\ g_{i-m,j-n} & \text{if } (i - m, j - n) \in \underline{k} \times \underline{l}; \\ 0 & \text{otherwise.} \end{cases}$$

A morphism $f: m \to n$ in the category $\mathbf{Mat}(\mathbb{F})$ can also be viewed as a m-by-n matrix

$$M_f = (f_{i,j})_{i \in \underline{m}, j \in \underline{n}}.$$

With this interpretation, the matrix associated to the composition $g \circ f$ is given by the matrix product

$$M_{g \circ f} = M_f M_g,$$

while the tensor product of two morphisms $f \otimes g$ is represented by the block matrix

$$M_{f \otimes g} = \begin{pmatrix} M_f & 0 \\ 0 & M_g \end{pmatrix}.$$

Example 1.5 Let G be a group and $H \subset G$ a normal subgroup. Denote by

$$C(H) := \{g \in G \mid gh = hg \ \forall h \in H\}$$

the commutant of H in G (which is also a normal subgroup of G). We define a category $\mathcal{G}_{H,G}$ where objects are elements of the quotient group G/H and a morphism $f: xH \to yH$ is a pair $(u, v) \in yH \times xH$ modulo the equivalence relation $(u, v) \sim (uh, vh)$, $h \in H$, such that $uv^{-1} \in C(H)$. Notice that the latter element does not change under the equivalence relation. The composition of morphisms $f = (u, v): xH \to yH$ and $g = (p, q): yH \to zH$ is given by the formula

$$g \circ f = (p, q) \circ (u, v) = (pq^{-1}u, v): xH \to zH.$$

This formula is explicitly compatible with the equivalence relation and thus is well defined. Associativity of the composition is verified straightforwardly

$$(f \circ g) \circ h = ((p, q) \circ (s, t)) \circ (u, v) = (pq^{-1}s, t) \circ (u, v) = (pq^{-1}st^{-1}u, v)$$

$$= (p, q) \circ (st^{-1}u, v) = (p, q) \circ ((s, t) \circ (u, v)) = f \circ (g \circ h).$$

The identity morphism id_{xH} is represented by (x, x).

This is a strict monoidal category where $xH \otimes yH = xyH$, the unit object being $eH = H$, and for two morphisms $f = (u, v)$ and $g = (p, q)$ their tensor product is defined by the component-wise multiplication $f \otimes g = (up, vq)$, and the condition $uv^{-1} \in C(H)$ for a morphism (u, v) ensures compatibility of the tensor product with the composition

$$(a \otimes b) \circ (c \otimes d) = (a \circ c) \otimes (b \circ d).$$

Example 1.6 A special case of the previous example corresponds to a group G with the trivial subgroup $H = \{e\}$. In this case, the category $\mathcal{G}_{H,G}$ is given by G as the set objects and, for any pair of objects $x, y \in G$, there is exactly one morphism $(y, x): x \to y$. The group multiplication of G gives the monoidal structure of the category.   □

1.1.2 Braided Monoidal Categories

For any category $\mathcal{C}$, the *exchange* functor

$$\varsigma: \mathcal{C} \times \mathcal{C} \to \mathcal{C} \times \mathcal{C} \tag{1.13}$$

is defined by exchanging the components, that is

$$\varsigma(A, B) = (B, A), \quad \forall (A, B) \in \mathrm{Ob}(\mathcal{C} \times \mathcal{C}), \tag{1.14}$$

for objects and

$$\varsigma(f, g) = (g, f), \quad (f, g): (A, B) \to (C, D), \tag{1.15}$$

for arrows (morphisms).

Let $\mathcal{C}$ be now a monoidal category. We have two functors

$$\otimes: \mathcal{C} \times \mathcal{C} \to \mathcal{C}, \quad (A, B) \mapsto A \otimes B, \quad (f, g) \mapsto f \otimes g \tag{1.16}$$

and

$$\otimes^{\mathrm{op}} = \otimes \circ \varsigma: \mathcal{C} \times \mathcal{C} \to \mathcal{C}, \quad (A, B) \mapsto B \otimes A, \quad (f, g) \mapsto g \otimes f. \tag{1.17}$$

A *braiding* in a monoidal category $\mathcal{C}$ is a natural isomorphism

$$\beta: \otimes \to \otimes^{\mathrm{op}} \tag{1.18}$$

such that the following diagrams are commutative:

$$A \otimes (B \otimes C) \xrightarrow{\beta_{A,B\otimes C}} (B \otimes C) \otimes A \xrightarrow{\alpha_{B,A,C}} B \otimes (C \otimes A)$$

$$\uparrow{\alpha_{A,B,C}} \qquad\qquad\qquad id_B \otimes \beta_{A,C} \uparrow$$

$$(A \otimes B) \otimes C \xrightarrow{\beta_{A,B}\otimes id_C} (B \otimes A) \otimes C \xrightarrow{\alpha_{B,A,C}} B \otimes (A \otimes C) \qquad (1.19)$$

and

$$(A \otimes B) \otimes C \xrightarrow{\beta_{A\otimes B,C}} C \otimes (A \otimes B) \xrightarrow{\alpha_{C,A,B}^{-1}} (C \otimes A) \otimes B$$

$$\uparrow{\alpha_{A,B,C}^{-1}} \qquad\qquad\qquad \beta_{A,C}\otimes id_B \uparrow$$

$$A \otimes (B \otimes C) \xrightarrow{id_A \otimes \beta_{B,C}} A \otimes (C \otimes B) \xrightarrow{\alpha_{A,C,B}^{-1}} (A \otimes C) \otimes B \qquad (1.20)$$

A *braided monoidal category* is a monoidal category with a braiding.

In view of applications in knot theory, braided monoidal categories constitute a very important class of monoidal categories, and the quantum double construction for Hopf algebras of Chap. 5 implicitly gives rise to braided monoidal categories, see, for example, [19, 21, 41].

A *symmetric monoidal category* is a braided monoidal category where the braiding satisfies the conditions

$$\beta_{A,B}^{-1} = \beta_{B,A}, \quad \forall A, B \in \mathrm{Ob}\,\mathcal{C}. \qquad (1.21)$$

In this case, the braiding is called *symmetry* and denoted as σ.

Example 1.7 The category **Set** of sets, see Example 1.1, is a symmetric monoidal category where the symmetry is given by the exchange maps

$$\sigma_{X,Y}: X \times Y \to Y \times X, \quad (x, y) \mapsto (y, x). \qquad (1.22)$$

As we will see in Sect. 1.2, any group can be interpreted as an object of this symmetric monoidal category. □

Example 1.8 The category **Vect**$_\mathbb{F}$ of $\mathbb{F}$-vector spaces, see Example 1.2, is a symmetric monoidal category where the symmetry is given by the exchange maps extended by linearity

$$\sigma_{V,W}: V \otimes_\mathbb{F} W \to W \otimes_\mathbb{F} V, \quad x \otimes_\mathbb{F} y \mapsto y \otimes_\mathbb{F} x. \qquad (1.23)$$

Hopf algebras are objects of this category, and the symmetry enters in their definition in one of the defining properties, see Definition 1.6 of Sect. 1.4. □

Example 1.9 For any $q \in \mathbb{F}_{\neq 0}$, the category $\mathbf{Vect}_{\mathbb{F}}^{\mathbb{Z}}$ of $\mathbb{Z}$-graded $\mathbb{F}$-vector spaces, see Example 1.3, is a braided monoidal category with the braiding

$$\beta_{U,V} : U \otimes V \to V \otimes U \tag{1.24}$$

defined by

$$\beta_{U,V}(x \otimes_{\mathbb{F}} y) = q^{mn} y \otimes_{\mathbb{F}} x, \quad x \in U_m, \quad y \in V_n, \quad m, n \in \mathbb{Z}. \tag{1.25}$$

It is a symmetric monoidal category if $q^2 = 1$.   □

1.1.3   The Graphical Notation of String Diagrams

Throughout this book, we will find it convenient sometimes to use the graphical notation of *string diagrams*.

Let $\mathcal{C}$ be a category. To any morphism $f : X \to Y$ in $\mathcal{C}$, we associate a graphical picture

$$f =: \quad \begin{array}{c} Y \\ | \\ \boxed{f} \\ | \\ X \end{array} \ . \tag{1.26}$$

If $f : X \to Y$ and $g : Y \to Z$ are two composable morphisms, then their composition is described by the vertical concatenation of graphs

$$g{\circ}f \ = \ \begin{array}{c} Z \\ | \\ \boxed{g \circ f} \\ | \\ X \end{array} \ = \ \begin{array}{c} Z \\ | \\ \boxed{g} \\ | \\ \boxed{f} \\ | \\ X \end{array} \tag{1.27}$$

In particular, for the identity morphism id_X it is natural to use just a line

$$\mathrm{id}_X \ = \ \begin{array}{c} X \\ | \\ \boxed{\mathrm{id}_X} \\ | \\ X \end{array} \ =: \ \begin{array}{c} X \\ | \\ | \\ X \end{array} \ . \tag{1.28}$$

The string diagrams are especially useful in the case when $\mathcal{C}$ is a strict monoidal category, because the tensor (monoidal) product can be drawn by the horizontal juxtaposition. Namely, for two morphisms $f: X \to Y$ and $g: U \to V$, their tensor product $f \otimes g: X \otimes U \to Y \otimes V$ is drawn as follows:

$$f \otimes g \; = \; \boxed{f \otimes g} \; =: \; \boxed{f \otimes g} \; =: \; \boxed{f}\;\boxed{g}.$$

(1.29)

By taking into account the distinguished role of the identity object I, it is natural to associate to it the empty graph.

In this notation, for example, the commutative diagram (1.19) for a braiding, in the context of a strict monoidal category, corresponds to the following diagrammatic equality

$$\boxed{\beta_{A,B \otimes C}} \;\; = \;\; \boxed{\beta_{A,C}} \; \boxed{\beta_{A,B}}$$

(1.30)

and the graphical equality corresponding to the commutative diagram (1.20)

$$\boxed{\beta_{A \otimes B,C}} \;\; = \;\; \boxed{\beta_{A,C}} \; \boxed{\beta_{B,C}}.$$

(1.31)

These relations become intuitively natural and almost tautological, if one uses a notation for a braiding borrowed from knot diagrams

$$\boxed{\beta_{A,B}} \;\; =: \;\; \times$$

(1.32)

which, in the case of a symmetric monoidal category, can be further simplified by removing the indication of under-passing strands.

In the case of non-strict monoidal categories, the graphical calculus becomes less convenient because of the non-associativity of the tensor product. In this case, the horizontal juxtaposition is not enough so that one should provide an extra structure, for example, the relative distance between the vertical lines.

More systematic and detailed explanation of the graphical notation of string diagrams can be found in Chapter 2 of the book [42].

1.2 Groups in Terms of Structural Maps

At first glance, the formal definition of a Hopf algebra (to be given later in Sect. 1.4) looks neither simple nor intuitively motivated. For this reason, we start by reviewing the notion of a group which we reformulate by using the structural maps as the basic entities. Such a reformulation will make the definition of a Hopf algebra very natural, at least from the viewpoint of group theory.

Recall that a *group* is a set G where, for any two elements $g, h \in G$, there corresponds a unique element gh called the *product* of g and h, a distinguished element e called the *identity element*, and, for any element g, there corresponds a unique element g^{-1} called the *inverse element* such that the following axioms are satisfied:

$$\text{associativity}: \quad (fg)h = f(gh), \quad \forall f, g, h \in G, \tag{1.33}$$

$$\text{unitality}: \quad eg = ge = g, \quad \forall g \in G, \tag{1.34}$$

$$\text{invertibility}: \quad gg^{-1} = g^{-1}g = e, \quad \forall g \in G. \tag{1.35}$$

We formalize the definition of a group by introducing three *structural maps*:

$$\text{product} \quad \mu: G \times G \to G, \quad (g, h) \mapsto gh, \tag{1.36}$$

$$\text{unit} \quad \eta: \underline{1} \to G, \quad 0 \mapsto e, \tag{1.37}$$

$$\text{inverse} \quad S: G \to G, \quad g \mapsto g^{-1}. \tag{1.38}$$

By rewriting

$$(fg)h = \mu(fg, h) = \mu(\mu(f, g), h) = \mu((\mu \times \mathrm{id})(f, g, h))$$

$$= \mu \circ (\mu \times \mathrm{id})(f, g, h) \tag{1.39}$$

and

$$f(gh) = \mu(f, gh) = \mu(f, \mu(g, h)) = \mu((\mathrm{id} \times \mu)(f, g, h))$$
$$= \mu \circ (\mathrm{id} \times \mu)(f, g, h), \qquad (1.40)$$

we conclude that the associativity axiom is equivalent to the following equality for the product map

$$\mu \circ (\mu \times \mathrm{id}) = \mu \circ (\mathrm{id} \times \mu) \qquad (1.41)$$

which corresponds to the commutative diagram

$$\begin{CD}
G \times G \times G @>{\mu \times \mathrm{id}}>> G \times G \\
@V{\mathrm{id} \times \mu}VV @VV{\mu}V \\
G \times G @>>{\mu}> G
\end{CD} \qquad . \qquad (1.42)$$

Before going further with the unitality axiom, let us agree on the following convention.

Let $\underline{1} := \{0\}$ be the set consisting of one element denoted by 0. For any set X, the (cartesian) product sets $\underline{1} \times X$ and $X \times \underline{1}$ are identified with X through the obvious canonical bijections $(0, x) \mapsto x$ and $(x, 0) \mapsto x$ which allows us to have natural identifications $(0, x) = (x, 0) = x$. With this convention, for the unitality axiom, we have

$$e \cdot g = \mu(\eta(0), g) = \mu \circ (\eta \times \mathrm{id})(0, g) = \mu \circ (\eta \times \mathrm{id})(g) \qquad (1.43)$$

and

$$g \cdot e = \mu(g, \eta(0)) = \mu \circ (\mathrm{id} \times \eta)(g, 0) = \mu \circ (\mathrm{id} \times \eta)(g) \qquad (1.44)$$

so that the unitality axiom can be stated as the following equations for the structural maps

$$\mu \circ (\eta \times \mathrm{id}) = \mathrm{id} = \mu \circ (\mathrm{id} \times \eta) \qquad (1.45)$$

which correspond to the commutative diagram

$$\begin{CD}
\underline{1} \times G @= G @= G \times \underline{1} \\
@V{\eta \times \mathrm{id}}VV @VV{\mathrm{id}}V @VV{\mathrm{id} \times \eta}V \\
G \times G @>{\mu}> G @<{\mu}< G \times G
\end{CD} \qquad . \qquad (1.46)$$

In order to describe the invertibility axiom in terms of equations for the structural maps, we need to use two other maps which are canonically defined for any set X. These are the *diagonal* or *coproduct* map

$$\Delta: X \to X \times X, \quad x \mapsto (x, x), \quad \forall x \in X, \tag{1.47}$$

and the *counit* map

$$\epsilon: X \to \underline{1}, \quad x \mapsto 0, \quad \forall x \in X. \tag{1.48}$$

These names come from the fact that they are similar to the product and the unit maps of a group in the following sense.

Definition 1.2 A commutative diagram Γ is called a *categorial* or *diagrammatic dual* of another commutative diagram Γ', if Γ can be obtained from Γ' by reversing all arrows and relabelling the objects.

The diagonal map satisfies the equality

$$(\mathrm{id} \times \Delta) \circ \Delta = (\Delta \times \mathrm{id}) \circ \Delta \tag{1.49}$$

corresponding to the commutative diagram

$$
\begin{array}{ccc}
X \times X \times X & \xleftarrow{\Delta \times \mathrm{id}} & X \times X \\
{\scriptstyle \mathrm{id} \times \Delta} \uparrow & & \uparrow {\scriptstyle \Delta} \\
X \times X & \xleftarrow{\quad \Delta \quad} & X
\end{array}
\tag{1.50}
$$

which is the categorial dual of the commutative diagram (1.42) corresponding to equality (1.41). For this reason, equality (1.49) is called the *coassociativity* property.

The counit map enters the *counitality* equalities

$$(\epsilon \times \mathrm{id}) \circ \Delta = \mathrm{id} = (\mathrm{id} \times \epsilon) \circ \Delta. \tag{1.51}$$

corresponding to the commutative diagram

$$
\begin{array}{ccccc}
\underline{1} \times X & =\!=\!= & X & =\!=\!= & X \times \underline{1} \\
{\scriptstyle \epsilon \times \mathrm{id}} \uparrow & & \uparrow {\scriptstyle \mathrm{id}} & & \uparrow {\scriptstyle \mathrm{id} \times \epsilon} \\
X \times X & \xleftarrow{\Delta} & X & \xrightarrow{\Delta} & X \times X
\end{array}
\tag{1.52}
$$

which is the categorial dual of the commutative diagram (1.46) corresponding to equalities (1.45) expressing the unitality axiom.

In order to express the invertibility axiom in terms of equations for structural maps, we write

$$gg^{-1} = \mu(g, g^{-1}) = \mu \circ (\mathrm{id} \times S)(g, g) = \mu \circ (\mathrm{id} \times S) \circ \Delta(g), \tag{1.53}$$

$$g^{-1}g = \mu(g^{-1}, g) = \mu \circ (S \times \mathrm{id})(g, g) = \mu \circ (S \times \mathrm{id}) \circ \Delta(g), \tag{1.54}$$

and

$$e = \eta(0) = \eta(\epsilon(g)) = \eta \circ \epsilon(g). \tag{1.55}$$

Thus, the invertibility axiom is equivalent to the equations

$$\mu \circ (\mathrm{id} \times S) \circ \Delta = \eta \circ \epsilon = \mu \circ (S \times \mathrm{id}) \circ \Delta \tag{1.56}$$

corresponding to the commutative diagram

$$
\begin{array}{ccc}
G \times G & \xrightarrow{\mathrm{id} \times S} & G \times G \\
{\scriptstyle \Delta}\uparrow & & \downarrow{\scriptstyle \mu} \\
G \xrightarrow{\;\epsilon\;} 1 & \xrightarrow{\;\eta\;} & G \\
{\scriptstyle \Delta}\downarrow & & \uparrow{\scriptstyle \mu} \\
G \times G & \xrightarrow{S \times \mathrm{id}} & G \times G
\end{array}
\tag{1.57}
$$

which is the categorial dual of itself.

Finally, by using the canonical exchange map

$$\sigma = \sigma_{G,G} : G \times G \to G \times G, \quad (x, y) \mapsto (y, x), \tag{1.58}$$

we remark that the product and the coproduct satisfy the *compatibility* equality

$$(\mu \times \mu) \circ (\mathrm{id} \times \sigma \times \mathrm{id}) \circ (\Delta \times \Delta) = \Delta \circ \mu : G \times G \to G \times G \tag{1.59}$$

corresponding to the commutative diagram

$$
\begin{array}{ccc}
G \times G \times G \times G & \xrightarrow{\mathrm{id} \times \sigma \times \mathrm{id}} & G \times G \times G \times G \\
{\scriptstyle \Delta \times \Delta}\uparrow & & \downarrow{\scriptstyle \mu \times \mu} \\
G \times G \xrightarrow{\;\mu\;} & G \xrightarrow{\;\Delta\;} & G \times G
\end{array}
\tag{1.60}
$$

which is also the categorial dual of itself. Moreover, identity (1.59) holds even if G is replaced by any set X and the product μ by any binary operation $f : X \times X \to X$. Indeed, for any $(x, y) \in X \times X$, we have

$$(f \times f) \circ (\mathrm{id} \times \sigma \times \mathrm{id}) \circ (\Delta \times \Delta)(x, y)$$

$$= (f \times f) \circ (\mathrm{id} \times \sigma \times \mathrm{id})(x, x, y, y) = (f \times f)(x, y, x, y) = (f(x, y), f(x, y))$$

$$= \Delta(f(x, y)) = \Delta \circ f(x, y). \qquad (1.61)$$

As the matter of fact, this calculation reflects an elementary general property of the cartesian symmetric monoidal category of sets which will be described in Sect. 1.3.

1.2.1 The Structural Maps of a Group in Graphical Notation

We are ready now to use the graphical notation of string diagrams introduced in Sect. 1.1.3 to rewrite the definition of a group. The monoidal category we are working in is the symmetric monoidal category **Set** of sets with the tensor product specified by the Cartesian product of sets, see Example 1.7.

Let us introduce the following graphical notation for the structural maps of a group (all lines correspond to the underlying set of the group G and the singleton $\underline{1}$ carries no line):

$$\text{product} \quad \mu =: \boxed{\mu} =: \bigwedge \qquad (1.62)$$

$$\text{coproduct} \quad \Delta =: \boxed{\Delta} =: \bigvee \qquad (1.63)$$

$$\text{unit} \quad \eta =: \boxed{\eta} =: \overset{\circ}{|} \qquad (1.64)$$

$$\text{counit} \quad \epsilon =: \boxed{\epsilon} =: \overset{|}{\bullet} \qquad (1.65)$$

$$\text{inverse or antipode} \quad S =: \boxed{S} =: \overset{|}{\underset{|}{\square}} \qquad (1.66)$$

$$\text{exchange or symmetry} \quad \sigma =: \boxed{\sigma} =: \times . \qquad (1.67)$$

For the inverse map in (1.66) we also put the term "antipode" in anticipation of its counterpart in the case of Hopf algebras, see Definition 1.6 of Sect. 1.4.

Recall from Sect. 1.1.3 that the composition of maps corresponds to vertical concatenation of the corresponding graphical objects, while the Cartesian product corresponds to horizontal juxtaposition. With this notation, the structural equations of a group take the following form:

associativity: (1.68)

unitality: (1.69)

coassociativity: (1.70)

counitality: (1.71)

invertibility: (1.72)

compatibility: (1.73)

Remark 1.1 A motivational idea behind the definition of a Hopf algebra is to think of these diagrams in the context of other symmetric monoidal categories. The corresponding realizations are called *group objects*. In particular, as we will see later in Sect. 1.4, a *Hopf algebra* can be identified as a group object in the symmetric monoidal category of vector spaces with the tensor product as the monoidal product, see Example 1.8.

1.3 Monoids and Comonoids

Given an algebraic notion, for example a group, it is often useful and instructive to consider other structures obtained from the initial one by dropping some of the defining properties/axioms. In the definition of a group, if we remove the inverse map together with the invertibility axiom, then we obtain the notion of a monoid.

Definition 1.3 A *monoid* is a set M together with two maps

$$\mu : M \times M \to M, \quad \eta : \underline{1} \to M, \quad (1.74)$$

respectively called *product* and *unit*, which satisfy the associativity axiom (1.41) and the unitality axiom (1.45).

Exercise 1.1 (Uniqueness of Inverses) An element $x \in M$ of a monoid M is called *invertible* if there exists an element $y \in M$, called *inverse* of x, such that $\mu(x, y) = \mu(y, x) = \eta(0)$. Show that any invertible element x admits a unique inverse.

Definition 1.4 Let $M = (M, \mu_M, \eta_M)$ and $N = (N, \mu_N, \eta_N)$ be two monoids. A map $f : M \to N$ is called *morphism of monoids* if it commutes with the structural maps in the sense of the relations

$$f \circ \mu_M = \mu_N \circ (f \times f), \quad f \circ \eta_M = \eta_N. \tag{1.75}$$

The notion of a *comonoid* is obtained by taking the categorial dual of the notion of a monoid, i.e. by reversing all arrows in the definition of a monoid in terms of commutative diagrams, see Definition 1.2.

Definition 1.5 A *comonoid* is a set C provided with two maps

$$\Delta : C \to C \times C, \quad \epsilon : C \to \underline{1}, \tag{1.76}$$

called *coproduct* and *counit*, which satisfy the coassociativity axiom (1.49) and the counitality axiom (1.51).

Exercise 1.2 Give a definition of a morphism of comonoids.

The following proposition shows that the notion of a comonoid is not particularly meaningful in a set-theoretic context.

Proposition 1.1 *Any set admits a unique comonoid structure and any map between two sets is a morphism of comonoids.*

Proof Notice that the counit map of any comonoid, being a map to the singleton, is uniquely fixed, and it is thus uniquely defined also for any set. Moreover, it is easily seen that any set X is a comonoid with the diagonal map as the coproduct and the map to the singleton as the counit. The formal proof is identical to the case of groups, see Eqs. (1.49)–(1.52). Let us show that there are no other comonoids.

Let $C = (C, \Delta, \epsilon)$ be a comonoid. Then the coproduct Δ corresponds to two maps $\alpha, \beta : C \to C$ defined by

$$\Delta(x) = (\alpha(x), \beta(x)). \tag{1.77}$$

Substituting this into the counitality axiom (1.51), we obtain

$$(\epsilon \times \mathrm{id}) \circ \Delta(x) = (\epsilon \times \mathrm{id})(\alpha(x), \beta(x))$$

$$= (\epsilon(\alpha(x)), \beta(x)) = (0, \beta(x)) = \beta(x) = x, \tag{1.78}$$

and

$$(\mathrm{id} \times \epsilon) \circ \Delta(x) = (\mathrm{id} \times \epsilon)(\alpha(x), \beta(x))$$
$$= (\alpha(x), \epsilon(\beta(x))) = (\alpha(x), 0) = \alpha(x) = x \qquad (1.79)$$

where we use the convention on the equality for the canonical identifications

$$X \times \underline{1} \simeq X \simeq \underline{1} \times X.$$

Thus, the coproduct is necessarily the diagonal map $\Delta(x) = (x, x)$.

Finally, any map between two sets $f : X \to Y$ enters the obvious commutative diagrams

$$
\begin{array}{ccc}
X & \xrightarrow{\;f\;} & Y \\
\big\downarrow{\scriptstyle \Delta} & & \big\downarrow{\scriptstyle \Delta} \\
X \times X & \xrightarrow{\;f \times f\;} & Y \times Y
\end{array}
\qquad \text{and} \qquad
\begin{array}{ccc}
X & \xrightarrow{\;f\;} & Y \\
& \underset{\epsilon}{\searrow}\;\;\underset{\epsilon}{\swarrow} & \\
& 1 &
\end{array}
\qquad (1.80)
$$

which mean that f is a morphism of comonoids. $\square$

1.4 Hopf Algebras

In this section, we introduce the central object of this book, a Hopf algebra. The definition that follows is motivated by the notion of a group which we reformulated in Sect. 1.2 in terms of structural maps. As the notions of a monoid or/and a comonoid are the results of dropping some of the structural maps and axioms from the definition of a group, in the subsequent sections of this chapter, we also introduce the analogous notions of an algebra (Sect. 1.6) and a coalgebra (Sect. 1.7) by dropping the corresponding structural maps and axioms from the definition of a Hopf algebra. There is also a notion of a bialgebra (Sect. 1.10), which in the set-theoretical context, corresponds to a monoid, but in the context of vector spaces, it is not true that every algebra (or coalgebra) is a bialgebra, though any Hopf algebra is a bialgebra. The reason for this difference comes from the fact that any set is canonically a comonoid in a unique way, as we have seen in the previous section, see Proposition 1.1, while a given vector space can admit many structures of a coalgebra, and, for a given algebra, there could be different possibilities, for example, non of available coalgebra structures on the underlying vector space can be compatible with the algebra structure, etc.

In what follows, we let $\mathbb{F}$ denote a field, write $\otimes$ instead of $\otimes_{\mathbb{F}}$ and omit the composition symbol in the case of linear maps.

Definition 1.6 A *Hopf algebra* (over a field $\mathbb{F}$) is a $\mathbb{F}$-vector space H of strictly positive dimension together with the following five linear maps:

$$\text{product} \quad \mu: H \otimes H \rightarrow H, \tag{1.81}$$

$$\text{coproduct} \quad \Delta: H \rightarrow H \otimes H, \tag{1.82}$$

$$\text{unit} \quad \eta: \mathbb{F} \rightarrow H, \tag{1.83}$$

$$\text{counit} \quad \epsilon: H \rightarrow \mathbb{F}, \tag{1.84}$$

$$\text{antipode} \quad S: H \rightarrow H, \tag{1.85}$$

which satisfy the following equations (axioms):

$$\text{associativity}: \quad \mu(\mu \otimes \mathrm{id}_H) = \mu(\mathrm{id}_H \otimes \mu), \tag{1.86}$$

$$\text{unitality}: \quad \mu(\eta \otimes \mathrm{id}_H) = \mathrm{id}_H = \mu(\mathrm{id}_H \otimes \eta), \tag{1.87}$$

$$\text{coassociativity}: \quad (\mathrm{id}_H \otimes \Delta)\Delta = (\Delta \otimes \mathrm{id}_H)\Delta, \tag{1.88}$$

$$\text{counitality}: \quad (\epsilon \otimes \mathrm{id}_H)\Delta = \mathrm{id}_H = (\mathrm{id}_H \otimes \epsilon)\Delta, \tag{1.89}$$

$$\text{invertibility}: \quad \mu(\mathrm{id}_H \otimes S)\Delta = \eta\epsilon = \mu(S \otimes \mathrm{id}_H)\Delta, \tag{1.90}$$

$$\text{compatibility}: \quad (\mu \otimes \mu)(\mathrm{id}_H \otimes \sigma \otimes \mathrm{id}_H)(\Delta \otimes \Delta) = \Delta\mu, \tag{1.91}$$

where the symmetry map $\sigma = \sigma_{H,H}: H \otimes H \rightarrow H \otimes H$ acts by $\sigma(x \otimes y) = y \otimes x$.

Remark 1.2 The list of axioms (1.86)–(1.91) exactly corresponds to the list of graphical relations (1.68)–(1.73), and we will often use the same graphical notation in this new context with the replacements:

$$\text{sets} \mapsto \text{vector spaces}$$
$$\text{set theoretical maps} \mapsto \text{linear maps}$$
$$\text{the singleton } \underline{1} = \{0\} \mapsto \text{the base field } \mathbb{F} \text{ (a 1-dimensional vector space)}$$
$$\text{the cartesian product} \mapsto \text{the tensor product.}$$

Definition 1.7 Let

$$H = (H, \mu_H, \eta_H, \Delta_H, \epsilon_H, S_H) \quad \text{and} \quad L = (L, \mu_L, \eta_L, \Delta_L, \epsilon_L, S_L)$$

be two Hopf algebras. A linear map $f : H \to L$ is called a *morphism of Hopf algebras* or a *Hopf algebra morphism* if it commutes with all the structural maps, that is the following diagrams are commutative:

$$
\begin{array}{ccc}
H \otimes H & \xrightarrow{f \otimes f} & L \otimes L \\
\mu_H \downarrow & & \downarrow \mu_L \\
H & \xrightarrow{\quad f \quad} & L
\end{array}
\qquad
\begin{array}{ccc}
H & \xrightarrow{\quad f \quad} & L \\
& \eta_H \nwarrow \quad \nearrow \eta_L & \\
& \mathbb{F} &
\end{array}
\qquad (1.92)
$$

$$
\begin{array}{ccc}
H \otimes H & \xrightarrow{f \otimes f} & L \otimes L \\
\Delta_H \uparrow & & \uparrow \Delta_L \\
H & \xrightarrow{\quad f \quad} & L
\end{array}
\qquad
\begin{array}{ccc}
H & \xrightarrow{\quad f \quad} & L \\
& \epsilon_H \searrow \quad \swarrow \epsilon_L & \\
& \mathbb{F} &
\end{array}
\qquad (1.93)
$$

$$
\begin{array}{ccc}
H & \xrightarrow{\quad f \quad} & L \\
s_H \downarrow & & \downarrow s_L \\
H & \xrightarrow{\quad f \quad} & L
\end{array}
\qquad (1.94)
$$

These commutative diagrams correspond to the following equalities between linear maps

$$\mu_L(f \otimes f) = f\mu_H, \quad f\eta_H = \eta_L, \qquad (1.95)$$

$$(f \otimes f)\Delta_H = \Delta_L f, \quad \epsilon_L f = \epsilon_H, \qquad (1.96)$$

$$S_L f = f S_H. \qquad (1.97)$$

Remark 1.3 In any Hopf algebra, we have the inequality $\eta \neq 0$ as otherwise the Hopf algebra would be zero-dimensional.

1.5 Group Algebras as Hopf Algebras

In this section we consider a class of examples of Hopf algebras coming from groups. Given the fact that we have motivated the definition of a Hopf algebra by considering the definition of a group, it is not very surprising that the two notions are related.

Definition 1.8 For any set X, we denote by $\delta_{a,b}$ the *Kronecker delta* function which is the characteristic function $\chi_{\Delta(X)} \colon X \times X \to \{0, 1\}$ of the diagonal $\Delta(X)$ in $X \times X$, i.e.

$$\delta_{a,b} = \chi_{\Delta(X)}(a, b) = \begin{cases} 1 \text{ if } a = b, \\ 0 \text{ if } a \neq b. \end{cases} \tag{1.98}$$

Definition 1.9 Let X be a set. The vector space of all maps $f \colon X \to \mathbb{F}$ of finite support, that is a map that takes all but finitely many values zero, is called the *vector space freely generated by* X, and it is denoted as $\mathbb{F}[X]$. A natural linear basis in $\mathbb{F}[X]$ is given by the set of single element characteristic functions $\{\chi_a\}_{a \in X}$ defined by

$$\chi_a(b) = \delta_{a,b}, \quad \forall (a, b) \in X^2. \tag{1.99}$$

Remark 1.4 The vector space $\mathbb{F}[X]$ can also be described as the direct sum of a family of 1-dimensional vector spaces $\mathbb{F}$ indexed by the set X:

$$\mathbb{F}[X] = \bigoplus_{x \in X} \mathbb{F}, \tag{1.100}$$

see Definition 1.1.

For any group G, let $\mathbb{F}[G]$ be the vector space freely generated by the set G. We define the product

$$\mu \colon \mathbb{F}[G] \otimes \mathbb{F}[G] \to \mathbb{F}[G], \quad (\mu(f \otimes g))(a) = \sum_{b \in G} f(b)g(b^{-1}a), \quad \forall a \in G, \tag{1.101}$$

the unit

$$\eta \colon \mathbb{F} \to \mathbb{F}[G], \quad (\eta 1)(a) = \delta_{e,a}, \quad \forall a \in G, \tag{1.102}$$

the coproduct

$$\Delta \colon \mathbb{F}[G] \to \mathbb{F}[G] \otimes \mathbb{F}[G], \quad (\Delta f)(a, b) = \delta_{a,b} f(a), \quad \forall (a, b) \in G^2, \tag{1.103}$$

the counit

$$\epsilon f = \sum_{a \in G} f(a), \tag{1.104}$$

and the antipode

$$(Sf)(a) = f(a^{-1}), \quad \forall a \in G. \tag{1.105}$$

Remark that in the definition of the product (1.101), the sum is finite due to the fact that the functions f and g are finitely supported. For the same reason, the function $\mu(f \otimes g)$ is also finitely supported. Indeed if $\mathrm{Supp}_f \subset G$ is the support of $f \in \mathbb{F}[G]$, then we have the inclusion

$$\mathrm{Supp}_{\mu(f \otimes g)} \subset \bigcup_{a \in \mathrm{Supp}_f} a \, \mathrm{Supp}_g . \tag{1.106}$$

With respect to the natural basis of single element characteristic functions $\{\chi_a\}_{a \in G}$, the structural maps take the form

$$\mu(\chi_a \otimes \chi_b) = \chi_{ab}, \quad \Delta \chi_a = \chi_a \otimes \chi_a, \tag{1.107}$$

$$\eta 1 = \chi_e, \quad \epsilon \chi_a = 1, \quad S \chi_a = \chi_{a^{-1}}. \tag{1.108}$$

Exercise 1.3 Show that the data $(\mathbb{F}[G], \mu, \eta, \Delta, \epsilon, S)$ satisfy the Hopf algebra axioms.

1.6   Algebras

Here we introduce the notion of an algebra by dropping some of the data in the definition of a Hopf algebra, namely we leave only the product and the unit as structural maps and impose on them the axioms of associativity and unitality corresponding to diagrammatic equations (1.68) and (1.69). An algebra is a monoidal object in the monoidal category of vector spaces with the tensor product as the monoidal product.

Definition 1.10 An *algebra* over a field $\mathbb{F}$ or $\mathbb{F}$-*algebra* is a triple (A, μ, η) consisting of a $\mathbb{F}$-vector space A, a linear map $\mu \colon A \otimes A \to A$ called *product*, and a linear map $\eta \colon \mathbb{F} \to A$ called *unit* such that

$$\mu(\mu \otimes \mathrm{id}_A) = \mu(\mathrm{id}_A \otimes \mu) \tag{1.109}$$

and

$$\mu(\eta \otimes \mathrm{id}_A) = \mu(\mathrm{id}_A \otimes \eta) = \mathrm{id}_A. \tag{1.110}$$

Example 1.10 As Eqs. (1.109) and (1.110) coincide respectively with Eqs. (1.86) and (1.87), any Hopf algebra is an algebra, if we keep the product and the unit and forget about all other structural maps.   □

Example 1.11 Let V be a vector space. Then, the vector space $\text{End}(V)$ of all endomorphisms of V is an algebra with the product

$$\mu(f \otimes g) = fg, \quad \forall (f, g) \in (\text{End}(V))^2, \tag{1.111}$$

and the unit

$$\eta 1 = \text{id}_V . \tag{1.112}$$

In particular, the base field $\mathbb{F} \simeq \text{End}(\mathbb{F})$ is an algebra.   □

Definition 1.11 Let $A = (A, \mu_A, \eta_A)$ and $B = (B, \mu_B, \eta_B)$ be two algebras. A linear map $f \colon A \to B$ is called a *morphism of algebras* or an *algebra morphism* if it commutes with the structural maps in the sense of the equations

$$f \mu_A = \mu_B (f \otimes f) \tag{1.113}$$

and

$$f \eta_A = \eta_B. \tag{1.114}$$

Definition 1.12 The *opposite product* of an algebra $A := (A, \mu, \eta)$ is the linear map μ^{op} obtained by composing the product with the exchange map,

$$\mu^{\text{op}} := \mu \sigma_{A,A} = \boxed{\mu} : A \otimes A \to A, \quad x \otimes y \mapsto \mu(y \otimes x). \tag{1.115}$$

The algebra A is called *commutative* if the opposite product coincides with the product, $\mu^{\text{op}} = \mu$.

Exercise 1.4 Show that if $A := (A, \mu, \eta)$ is an algebra then $A^{\text{op}} := (A, \mu^{\text{op}}, \eta)$ is also an algebra.

Definition 1.13 Let $A_1 = (A_1, \mu_1, \eta_1)$ and $A_2 = (A_2, \mu_2, \eta_2)$ be two algebras. The *tensor product* of A_1 and A_2 is the algebra

$$(A_1 \otimes A_2, (\mu_1 \otimes \mu_2)(\text{id}_{A_1} \otimes \sigma_{A_2,A_1} \otimes \text{id}_{A_2}), \eta_1 \otimes \eta_2) \tag{1.116}$$

or graphically

$$\boxed{\mu_{A_1 \otimes A_2}} = \boxed{\mu_1}\ \boxed{\mu_2} \quad \text{and} \quad \boxed{\eta_{A_1 \otimes A_2}} = \boxed{\eta_1}\ \boxed{\eta_2} \tag{1.117}$$

where the thin lines correspond to A_1 and thick lines to A_2, and we implicitly identify $\mathbb{F} \otimes \mathbb{F}$ with $\mathbb{F}$.

Exercise 1.5 Let $A = (A, \mu, \eta)$ be an algebra. Show that the unit $\eta \colon \mathbb{F} \to A$ is always a morphism of algebras, while the product $\mu \colon A \otimes A \to A$ is a morphism of algebras if and only if A is commutative.

1.6.1 Iterated Products

Let $A = (A, \mu, \eta)$ be an $\mathbb{F}$-algebra. In calculations, it is the common practice to write just xy instead of $\mu(x \otimes y)$. In particular, as the associativity axiom (1.109) implies that $(xy)z = x(yz)$, one can just write xyz without any ambiguity. Graphically, this means that we can use multivalent vertices:

$$\includegraphics{} = \includegraphics{} =: \includegraphics{} . \tag{1.118}$$

This can be formalised by introducing the set of iterated products

$$\{\mu^{(m)} \colon A^{\otimes m} \to A\}_{m \in \omega} \tag{1.119}$$

defined recursively as follows:

$$\mu^{(m)} := \mu(\mu^{(m-1)} \otimes \mathrm{id}_A), \quad \mu^{(0)} := \eta, \tag{1.120}$$

so that, in particular, we have

$$\mu^{(1)} = \mathrm{id}_A, \quad \mu^{(2)} = \mu. \tag{1.121}$$

The n-th iterated product $\mu^{(n)}$ graphically can be represented by any binary tree with n inputs and one output, because the associativity of the product allows to ensure that any such tree gives one and the same linear map which we denote by a multivalent vertex.

Exercise 1.6 Prove that

$$\mu^{(k_1 + \cdots + k_m)} = \mu^{(m)}\left(\mu^{(k_1)} \otimes \cdots \otimes \mu^{(k_m)}\right), \quad \forall (k_1, \ldots, k_m) \in \omega^m. \tag{1.122}$$

Exercise 1.7 Let $f \colon A \to B$ be an algebra morphism. Prove that

$$\mu_B^{(m)} f^{\otimes m} = f \mu_A^{(m)}, \quad \forall m \in \omega. \tag{1.123}$$

1.6.2   Modules

In the context of vector spaces, the notion of a module over an algebra corresponds to an M-set in the set-theoretical context, that is a set on which a monoid M acts.

Definition 1.14 Let $A = (A, \mu, \eta)$ be an algebra over a field $\mathbb{F}$. A *left module* over A (or simply a *left A-module*) is a $\mathbb{F}$-vector space V together with a linear map

$$\lambda \colon A \otimes V \to V \tag{1.124}$$

such that the diagrams

$$\tag{1.125}$$

are commutative. In terms of our graphical notation, the commutative diagrams (1.125) correspond to the equations

$$\tag{1.126}$$

where the thick lines correspond to V and thin lines to A.

Remark 1.5 For two vector spaces X and Y, let $L(X, Y)$ be the set of all linear maps from X to Y. The natural bijection between two sets of linear maps

$$L(A \otimes V, V) \simeq L(A, \mathrm{End}(V)) \tag{1.127}$$

descents to a natural bijection between the sets of left A-module structures on V and algebra morphisms from A to $\mathrm{End}(V)$. For this reason, a left A-module structure on a vector space V is often called *representation* of A in V.

Exercise 1.8 Give a definition of a right module over an algebra A.

1.7 Coalgebras

The notion of a coalgebra is the categorial dual of that of an algebra in the sense that the commutative diagrams expressing the defining properties of an algebra and coalgebra are related through the categorial duality, see Definition 1.2. The definition of a coalgebra is obtained by dropping all the structural maps in the definition of a Hopf algebra, apart from the coproduct and the counit, and by keeping the axioms of coassociativity and counitality. These axioms correspond to two diagrammatic equations (1.70) and (1.71).

Definition 1.15 A *coalgebra* over a field $\mathbb{F}$ or a $\mathbb{F}$-*coalgebra* is a triple (C, Δ, ϵ) consisting of a $\mathbb{F}$-vector space C, a linear map $\Delta \colon C \to C \otimes C$ called *coproduct*, and a linear map $\epsilon \colon C \to \mathbb{F}$ called *counit* such that

$$(\Delta \otimes \mathrm{id}_C)\Delta = (\mathrm{id}_C \otimes \Delta)\Delta \tag{1.128}$$

and

$$(\epsilon \otimes \mathrm{id}_C)\Delta = (\mathrm{id}_C \otimes \epsilon)\Delta = \mathrm{id}_C. \tag{1.129}$$

Example 1.12 As Eqs. (1.128) and (1.129) coincide respectively with Eqs. (1.88) and (1.89), any Hopf algebra is a coalgebra, if we keep the coproduct and the counit and forget about all other structural maps.   □

Example 1.13 For a finite non-empty set I, let $\mathbb{F}[I^2]$ be the $\mathbb{F}$-vector space freely generated by the set $I^2 = I \times I$. Then, $\mathbb{F}[I^2]$ is a coalgebra, if, for the natural linear basis $\{\chi_{(i,j)}\}_{(i,j)\in I^2}$ of $\mathbb{F}[I^2]$, we define a coproduct

$$\Delta \chi_{(i,j)} = \sum_{k\in I} \chi_{(i,k)} \otimes \chi_{(k,j)} \tag{1.130}$$

and a counit

$$\epsilon \chi_{(i,j)} = \delta_{i,j}. \tag{1.131}$$

Through the duality relation between algebras and coalgebras to be discussed later, this coalgebra is closely related to the endomorphism algebra $\mathrm{End}(V)$, see Example 1.11, associated to a vector space of dimension given by the cardinality $|I|$ of the set I. This algebra, in its turn, through a choice of a basis in V, becomes the algebra of square matrices of size $|I|$. For this reason, this coalgebra is called *matrix coalgebra*.   □

Example 1.14 The vector space $\mathbb{F}[\mathbb{Z}_{>0}]$ freely generated by the set of strictly positive integers $\mathbb{Z}_{>0}$ is a coalgebra with the coproduct

$$\Delta \chi_m = \sum_{a \in \mathrm{Div}(m)} \chi_a \otimes \chi_{m/a}, \tag{1.132}$$

where $\mathrm{Div}(m)$ is the set of all (positive) divisors of m, and the counit

$$\epsilon \chi_m = \delta_{1,m}. \tag{1.133}$$

This coalgebra will be called *Dirichlet coalgebra* because of its role in analytic number theory, see Example 1.17 in the next Sect. 1.8. □

Exercise 1.9 Give a definition of a morphism of coalgebras.

Definition 1.16 The *opposite coproduct* in a coalgebra $C := (C, \Delta, \epsilon)$ is the linear map Δ^{op} obtained by composing the coproduct with the exchange map

$$\Delta^{\mathrm{op}} := \sigma \Delta = \begin{matrix} \asymp \\ \boxed{\Delta} \\ | \end{matrix}. \tag{1.134}$$

The coalgebra C is called *cocommutative* if the opposite coproduct coincides with the coproduct, $\Delta^{\mathrm{op}} = \Delta$.

Exercise 1.10 Show that if $C = (C, \Delta, \epsilon)$ is a coalgebra, then $C^{\mathrm{cop}} := (C, \Delta^{\mathrm{op}}, \epsilon)$ is also a coalgebra.

The following definition is motivated by the behavior of the canonical basis elements of group (Hopf) algebras under the coproduct, see relations (1.107).

Definition 1.17 A non zero element g of a coalgebra is called *grouplike* if $\Delta g = g \otimes g$.

Exercise 1.11 Show that any set of grouplike elements of a coalgebra is linearly independent.

Exercise 1.12 Show that the matrix coalgebra of Example 1.13 contains a grouplike element only if it is 1-dimensional.

The following definition introduces the notion of an element of a coalgebra which can be viewed as simplest among non grouplike elements.

Definition 1.18 A non zero element x of a coalgebra is called *primitive* if

$$\Delta x = g \otimes x + x \otimes h$$

where g, h are grouplike elements.

Exercise 1.13 Find grouplike and primitive elements in the Dirichlet coalgebra of Example 1.14.

Definition 1.19 Let $C_1 = (C_1, \Delta_1, \epsilon_1)$ and $C_2 = (C_2, \Delta_2, \epsilon_2)$ be two coalgebras. The *tensor product* of C_1 and C_2 is the coalgebra

$$(C_1 \otimes C_2, (\mathrm{id}_{C_1} \otimes \sigma_{C_1,C_2} \otimes \mathrm{id}_{C_2})(\Delta_1 \otimes \Delta_2), \epsilon_1 \otimes \epsilon_2) \qquad (1.135)$$

or graphically

$$(1.136)$$

where the thin lines correspond to C_1 and thick lines to C_2.

1.7.1 Iterated Coproducts

Similarly to the case of algebras, due to the coassociativity property, it is convenient to use multivalent vertices in graphical representation of iterated coproducts:

$$(1.137)$$

Elements of the infinite set of all iterated coproducts

$$\{\Delta^{(m)} \colon C \to C^{\otimes m}\}_{m \in \omega} \qquad (1.138)$$

are defined recursively

$$\Delta^{(m)} := (\Delta^{(m-1)} \otimes \mathrm{id}_C)\Delta, \quad \Delta^{(0)} = \epsilon, \qquad (1.139)$$

so that, in particular, we have

$$\Delta^{(1)} = \mathrm{id}_C, \quad \Delta^{(2)} = \Delta. \qquad (1.140)$$

1.7.2   Sweedler's Sigma Notation for the Iterated Coproducts

Originally introduced in the book [39], Sweedler's *sigma notation* allows to write formally the coproduct of an element of a coalgebra in the form

$$\Delta x = \sum_{(x)} x_{(1)} \otimes x_{(2)} \tag{1.141}$$

where the meaning of the sum is that it is a finite sum of the form

$$\Delta x = \sum_{i=1}^{n} a_i \otimes b_i \tag{1.142}$$

where the number n and the elements a_i, b_i with $1 \le i \le n$ are determined non uniquely by x. The sigma notation allows to avoid mentioning the number n and the associated elements all together thus simplifying writing. More generally, one can use a similar notation also for iterated coproducts

$$\Delta^{(m)} x = \sum_{(x)} x_{(1)} \otimes x_{(2)} \otimes \cdots \otimes x_{(m)}, \quad \forall m \ge 2. \tag{1.143}$$

In this notation, for example, the equality $\Delta^{(3)}(x) = ((\Delta \otimes \mathrm{id}_C) \circ \Delta)(x)$ takes the form

$$\sum_{(x)} x_{(1)} \otimes x_{(2)} \otimes x_{(3)} = \sum_{(x)} \sum_{(x_{(1)})} x_{(1)(1)} \otimes x_{(1)(2)} \otimes x_{(2)}. \tag{1.144}$$

For examples and exercices on using the sigma notation, see the book [39].

1.7.3   The Fundamental Theorem of Coalgebras

Despite the fact that coalgebras are categorially dual objects to algebras, there is an important difference between them. Namely, there is no a conterpart for algebras of the following theorem.

Theorem 1.1 (The Fundamental Theorem of Coalgebras) *Let $C = (C, \Delta, \epsilon)$ be a coalgebra and $x \in C$. Then, there exists a finite dimensional sub-coalgebra $X \subset C$ containing x.*

Proof As the case $x = 0$ is trivial, we assume that $x \ne 0$.

Let $\{\alpha_i\}_{i \in I}$ and $\{\beta_j\}_{j \in J}$ be two non empty finite sets of linearly independent elements of C such that

$$\Delta^{(3)}x = \sum_{(i,j) \in I \times J} \alpha_i \otimes x_{i,j} \otimes \beta_j \qquad (1.145)$$

and let $X \subset C$ be the vector subspace generated by the elements $\{x_{i,j}\}_{(i,j) \in I \times J}$. We have $\dim(X) \leq |I||J| < \infty$ and

$$x = (\epsilon \otimes \mathrm{id}_C \otimes \epsilon)\Delta^{(3)}x = \sum_{(i,j) \in I \times J} \epsilon(\alpha_i)\epsilon(\beta_j)x_{i,j} \in X. \qquad (1.146)$$

Let us show that X is a sub-coalgebra of C, that is $\Delta(X) \subset X \otimes X$.

We have the equalities

$$\Delta^{(4)}x = (\Delta \otimes \mathrm{id}_{C^{\otimes 2}})\Delta^{(3)}x = \sum_{(k,j) \in I \times J} (\Delta\alpha_k) \otimes x_{k,j} \otimes \beta_j \qquad (1.147)$$

$$\Delta^{(4)}x = (\mathrm{id}_C \otimes \Delta \otimes \mathrm{id}_C)\Delta^{(3)}x = \sum_{(i,j) \in I \times J} \alpha_i \otimes (\Delta x_{i,j}) \otimes \beta_j \qquad (1.148)$$

$$\Delta^{(4)}x = (\mathrm{id}_{C^{\otimes 2}} \otimes \Delta)\Delta^{(3)}x = \sum_{(i,l) \in I \times J} \alpha_i \otimes x_{i,l} \otimes (\Delta\beta_l). \qquad (1.149)$$

Comparing the right hand sides of (1.147) and (1.148) and using the linear independence of the family $\{\beta_j\}_{j \in J}$, we obtain the equalities

$$\sum_{k \in I} (\Delta\alpha_k) \otimes x_{k,j} = \sum_{i \in I} \alpha_i \otimes (\Delta x_{i,j}), \quad \forall j \in J, \qquad (1.150)$$

which, in their turn, due to the linear independence of the family $\{\alpha_i\}_{i \in I}$, imply that

$$\Delta\alpha_k = \sum_{i \in I} \alpha_i \otimes \alpha_{i,k}, \quad \forall k \in I, \qquad (1.151)$$

for some elements $\{\alpha_{i,k}\}_{i,k \in I} \subset C$ and

$$\Delta x_{i,j} = \sum_{k \in I} \alpha_{i,k} \otimes x_{k,j} \in C \otimes X, \quad \forall(i, j) \in I \times J. \qquad (1.152)$$

By a similar reasoning, comparing the right hand sides of (1.149) and (1.148), we obtain

$$\Delta \beta_l = \sum_{j \in I} \beta_{l,j} \otimes \beta_j, \quad \forall l \in J, \tag{1.153}$$

for some elements $\{\beta_{l,j}\}_{l,j \in J} \subset C$ and

$$\Delta x_{i,j} = \sum_{l \in J} x_{i,l} \otimes \beta_{l,j} \in X \otimes C, \quad \forall (i, j) \in I \times J. \tag{1.154}$$

Finally, putting together (1.152) and (1.154), we conclude that

$$\Delta x_{i,j} \in X \otimes X, \quad \forall (i, j) \in I \times J \quad \Rightarrow \quad \Delta(X) \subset X \otimes X. \tag{1.155}$$

$\square$

The fundamental theorem of coalgebras allows to reduce many questions about general coalgebras to questions about finite-dimensional coalgebras. Notice also that the category of finite dimensional coalgebras is equivalent to the category of finite dimensional algebras in the sense that the dual vector space of a finite dimensional algebra is canonically a finite dimensional coalgebra and vice versa.

1.7.4 Comodules

The notion of a comodule over a coalgebra is the categorial dual to that of a module over an algebra in the sense of Definition 1.2.

Definition 1.20 Let $C = (C, \Delta, \epsilon)$ be a coalgebra over a field $\mathbb{F}$. A *right comodule* over C (or simply a *right C-comodule*) is a $\mathbb{F}$-vector space V together with a linear map

$$\delta : V \to V \otimes C \tag{1.156}$$

such that the diagrams

$$\tag{1.157}$$

are commutative which, in the graphical notation, correspond to the equations

$$\begin{array}{c}\text{(diagram)}\end{array} = \begin{array}{c}\text{(diagram)}\end{array} \quad \text{and} \quad \begin{array}{c}\text{(diagram)}\end{array} = \begin{array}{c}|\end{array}$$

(1.158)

where the thick lines correspond to V and thin lines to C.

Exercise 1.14 Give a definition of a left comodule over a coalgebra C.

Example 1.15 An obvious example of a C-comodule (both right and left ones) is the coalgebra C itself with $\delta = \Delta$. □

1.8 Convolution Algebras

The dual vector space of any coalgebra is canonically an algebra called the convolution algebra of a coalgebra. This is a special case of a more general convolution algebra associated to an algebra and a coalgebra.

Proposition-Definition 1.1 *Let A be an algebra and C a coalgebra. Then, the vector space $L(C, A)$ of linear maps from C to A is an algebra, called* convolution algebra, *with the product $\mu\colon L(C, A) \otimes L(C, A) \to L(C, A)$ defined by*

$$\mu(f \otimes g) =: f * g := \mu_A(f \otimes g)\Delta_C \tag{1.159}$$

or diagrammatically

$$\boxed{\mu(f \otimes g)} = \boxed{f * g} := \begin{array}{c}\mu_A \\ f \quad g \\ \Delta_C\end{array},$$

(1.160)

where the thick lines correspond to C and thin lines to A, and the unit $\eta\colon \mathbb{F} \to L(C, A)$ is defined by

$$\eta 1 = \eta_A \epsilon_C = \boxed{\eta_A \epsilon_C} = \begin{array}{c}\eta_A \\ \epsilon_C\end{array}.$$

(1.161)

Proof We verify the associativity property

$$(f * g) * h = \mu_A((f * g) \otimes h)\Delta_C = \mu_A((\mu_A(f \otimes g)\Delta_C) \otimes h)\Delta_C$$

$$= \mu_A(\mu_A \otimes \mathrm{id}_A)(f \otimes g \otimes h)(\Delta_C \otimes \mathrm{id}_C)\Delta_C = \mu_A^{(3)}(f \otimes g \otimes h)\Delta_C^{(3)}$$

$$= \mu_A(\mathrm{id}_A \otimes \mu_A)(f \otimes g \otimes h)(\mathrm{id}_C \otimes \Delta_C)\Delta_C = \mu_A(f \otimes (\mu_A(g \otimes h)\Delta_C))\Delta_C$$

$$= \mu_A(f \otimes (g * h))\Delta_C = f * (g * h) \qquad (1.162)$$

and the unitality property

$$f * (\eta 1) = \mu_A(f \otimes (\eta 1))\Delta_C = \mu_A(f \otimes (\eta_A \epsilon_C))\Delta_C$$

$$= \mu_A(\mathrm{id}_A \otimes \eta_A)f(\mathrm{id}_C \otimes \epsilon_C)\Delta_C = \mathrm{id}_A \, f \, \mathrm{id}_C = f \qquad (1.163)$$

and similarly for the product $(\eta 1) * f$. □

Remark 1.6 In order to illustrate the effectiveness of the graphical calculus of string diagrams in this context, here is the diagrammatic proof of the associativity of the convolution product (cf. (1.162)):

$$(1.164)$$

and the unitality property of the convolution product (cf. (1.163)):

where we are using the simplified notation for the structural maps of the algebra A and the coalgebra C.

As the base field $\mathbb{F}$ is canonically an algebra, see Example 1.11, a particular case of the convolution algebra $L(C, A)$ with $A = \mathbb{F}$ corresponds to an algebra structure on the dual vector space $C^* = L(C, \mathbb{F})$ of a coalgebra C given by the product

$$\langle f * g, x \rangle = \langle f \otimes g, \Delta_C x \rangle = \sum_{(x)} \langle f, x_{(1)} \rangle \langle g, x_{(2)} \rangle, \tag{1.165}$$

where $\langle \cdot, \cdot \rangle \colon C^* \times C \to \mathbb{F}$ is the evaluation map of a linear form on a vector, and the unit element $\eta 1 = \epsilon_C \in C^*$. This algebra is called the *convolution algebra of a coalgebra*.

Example 1.16 The convolution algebra of the matrix coalgebra from Example 1.13 is isomorphic to the algebra of n-by-n matrices where $n = |I|$ is the cardinality of the set I. It is also identified with the endomorphism algebra $\text{End}(\mathbb{F}^n)$.   □

Example 1.17 The convolution algebra of the Dirichlet coalgebra (see Example 1.14) is known as the *Dirichlet convolution algebra*. Its subalgebra of arithmetic functions plays an important role in analytic number theory, where the corresponding convolution product is called Dirichlet product or Dirichlet convolution, see, for example, Chapter 2 of the book [2].   □

Exercise 1.15 An element of the Dirichlet convolution algebra $f \in (\mathbb{F}[\mathbb{Z}_{>0}])^*$ is called *multiplicative* if $\langle f, \chi_{ab} \rangle = \langle f, \chi_a \rangle \langle f, \chi_b \rangle$ for all mutually prime pairs of positive integers $a, b \in \mathbb{Z}_{>0}$ and $\langle f, \chi_1 \rangle = 1$. Show that if $f, g \in (\mathbb{F}[\mathbb{Z}_{>0}])^*$ are multiplicative, then their convolution product $f * g$ is also a multiplicative element.

1.9 Some Properties of Hopf Algebras

For a Hopf algebra H, the invertibility axiom (1.90) is nothing else but the condition that the antipode is the inverse of the identity map id_H in the convolution algebra $\text{End}(H)$.

By the uniqueness of inverses, this means that a Hopf algebra cannot admit more than one antipode. Indeed, assuming that $\tilde{S}$ is another element of $\text{End}(H)$ satisfying the invertibility axiom, we write the associativity condition for the triple of elements $(\tilde{S}, \text{id}_H, S)$ in the convolution algebra $\text{End}(H)$:

$$(\tilde{S} * \text{id}_H) * S = \tilde{S} * (\text{id}_H * S) \Leftrightarrow (\eta_{\text{End}(H)} 1) * S = \tilde{S} * (\eta_{\text{End}(H)} 1) \Leftrightarrow S = \tilde{S}. \tag{1.166}$$

Remark 1.7 Definition 1.6 of a Hopf algebra differs from the standard definition(s) in the literature. Specifically, in Definition 1.6 we do not assume that the counit (respectively the unit) is an algebra (respectively a coalgebra) morphism. Below, we derive these properties from the axioms listed in Definition 1.6. These derivations are based on the interpretations of the product μ and the coproduct Δ as invertible elements of the convolution algebras $L(H \otimes H, H)$ and $L(H, H \otimes H)$, respectively.

Lemma 1.1 *In any Hopf algebra $H = (H, \mu, \eta, \Delta, \epsilon, S)$, the product μ (respectively the coproduct Δ) is an invertible element of the convolution algebra $L(H \otimes H, H)$ (respectively $L(H, H \otimes H)$) with the inverse*

$$\bar{\mu} := \mu^{\mathrm{op}}(S \otimes S) = \overset{\text{\tiny graphic}}{\underset{}{}} \quad \left(respectively \; \bar{\Delta} := (S \otimes S)\Delta^{\mathrm{op}} = \overset{\text{\tiny graphic}}{\underset{}{}}\right).$$

(1.167)

Here the opposite product and the opposite coproduct are defined by

$$\mu^{\mathrm{op}} := \mu \sigma_{H,H}, \quad \Delta^{\mathrm{op}} := \sigma_{H,H} \Delta.$$

(1.168)

Proof Here is a graphical proof of the fact that $\bar{\mu}$ is a right convolution inverse of μ:

(1.169)

where, in the second equality, we convert the three trivalent vertices corresponding to the product into a multivalent vertex corresponding to an iterated product; in the third equality, by using associativity of the product and properties of the symmetry, we "pulled out" appropriately chosen trivalent vertex from the multivalent vertex, and in the last three equalities, we use twice the invertibility axiom and once the unitality axiom.

The rest of the proof goes along the same type of graphical calculations. □

Proposition 1.2 *In any Hopf algebra, the counit (respectively unit) is a morphism of algebras (respectively coalgebras). This means that*

$$\epsilon \mu = \epsilon \otimes \epsilon \quad \Leftrightarrow \quad \overset{\text{\tiny graphic}}{\underset{}{}} = \overset{\text{\tiny graphic}}{\underset{}{}},$$

(1.170)

$$\Delta \eta = \eta \otimes \eta \quad \Leftrightarrow \quad \overset{\text{\tiny graphic}}{\underset{}{}} = \overset{\text{\tiny graphic}}{\underset{}{}},$$

(1.171)

$$\epsilon \eta = \mathrm{id}_{\mathbb{F}} \quad \Leftrightarrow \quad \overset{\text{\tiny graphic}}{\underset{}{}} = 1 \quad \Leftrightarrow \quad \epsilon \eta 1 = 1.$$

(1.172)

Proof The compatibility, the unitality and the counitality axioms imply that

$$(\eta\epsilon\mu) * \mu = \mu$$

in the convolution algebra $L(H \otimes H, H)$. As μ is an invertible element, we conclude that

$$\eta\epsilon\mu = \eta_{L(H\otimes H,H)}1 = \eta(\epsilon\otimes\epsilon) \quad \Rightarrow \quad \epsilon\mu = \epsilon\otimes\epsilon \quad \Rightarrow \quad \epsilon\eta = \mathrm{id}_{\mathbb{F}}. \quad (1.173)$$

By the duality symmetry, the compatibility, the unitality and the counitality axioms imply that

$$(\Delta\eta\epsilon) * \Delta = \Delta$$

in the convolution algebra $L(H, H \otimes H)$. As Δ is an invertible element, we conclude that

$$\Delta\eta\epsilon = \eta_{L(H,H\otimes H)}1 = (\eta \otimes \eta)\epsilon \quad \Rightarrow \quad \Delta\eta = \eta \otimes \eta. \quad (1.174)$$

$$\square$$

Exercise 1.16 Show that if $H := (H, \mu, \eta, \Delta, \epsilon, S)$ is a Hopf algebra, then

$$H^{\mathrm{op,cop}} := (H, \mu^{\mathrm{op}}, \eta, \Delta^{\mathrm{op}}, \epsilon, S) \quad (1.175)$$

is also a Hopf algebra.

Proposition 1.3 *In any Hopf algebra H, the antipode is a Hopf algebra morphism from H to $H^{\mathrm{op,cop}}$.*

Proof By Lemma 1.1, the convolution inverse of μ is the map $\bar{\mu}$ defined in (1.167). On the other hand, the composition $S\mu$ is also the convolutional inverse of μ as shows the following diagrammatic calculation:

$$\mu*(S\mu) = \qquad\qquad\qquad\qquad\qquad\qquad\qquad\qquad\qquad (1.176)$$

and likewise for the product $(S\mu) * \mu$. Thus, by uniqueness of inverses, we have the equality

$$S\mu = \mu^{\mathrm{op}}(S \otimes S) \quad (1.177)$$

and likewise

$$\Delta S = (S \otimes S)\Delta^{\mathrm{op}}. \tag{1.178}$$

To finish the proof, we check that

$$\epsilon S = = = = = \epsilon \tag{1.179}$$

and similarly

$$S\eta = \eta. \tag{1.180}$$

$\square$

1.10 Bialgebras

Bialgebras, like Hopf algebras, are categorially self-dual algebraic objects (in the sense of Definition 1.2) that carry compatible structures of an algebra and a coalgebra but without assuming the existence of the antipode.

Definition 1.21 A *bialgebra* is a tuple $(B, \mu, \eta, \Delta, \epsilon)$, where (B, μ, η) is an algebra, (B, Δ, ϵ) is a coalgebra, and the linear maps Δ and ϵ are algebra morphisms (or, equivalently, μ and η are coalgebra morphisms).

Exercise 1.17 Give a definition of a bialgebra morphism.

Remark 1.8 By forgetting the antipode, any Hopf algebra becomes a bialgebra if one keeps the property that the counit is a morphism of algebras. A bialgebra B originates in this way from a Hopf algebra if and only if the identity map id_B is invertible in the convolution algebra $\mathrm{End}(B)$ of endomorphisms of B.

Example 1.18 Let M be a monoid, i.e. a set with associative product and the unit element $e \in M$, see Definition 1.3. The monoid bialgebra is the vector space $\mathbb{F}[M]$ freely generated by the set M, where the structure maps are given in terms of the linear basis of characteristic functions of points $\{\chi_a\}_{a \in M}$ by the formulae

$$\mu(\chi_a \otimes \chi_b) = \chi_{ab}, \quad \eta 1 = \chi_e, \tag{1.181}$$

$$\Delta \chi_a = \chi_a \otimes \chi_a, \quad \epsilon \chi_a = 1. \tag{1.182}$$

These relations coincide with the relations (1.107) for group algebras, so that verification of the axioms follow the same line of reasoning as in the case of group algebras. □

Exercise 1.18 Show that a monoid bialgebra $\mathbb{F}[M]$ admits the structure of a Hopf algebra if and only if M is a group.

Chapter 2
Constructions of Algebras, Coalgebras, Bialgebras, and Hopf Algebras

Our main goal in this chapter is to show that Hopf algebras are abundant, and that there exist general methods of producing examples. We start by describing constructions of algebras which are common and well known. Then, we proceed with less well known constructions of coalgebras, bialgebras and Hopf algebras.

Coalgebras and algebras, being in categorial duality to each other, are closely related. For this reason, construction of an algebra also gives us a coalgebra. By using vector spaces as input in the case of algebras, and coalgebras in the case of bialgebras and Hopf algebras, we describe the constructions of free objects which have universal properties, and then realise all other objects as quotients of free objects.

2.1 Construction of Algebras

Any algebra can be constructed by using a vector space as an input. A special role is played by free algebras which, due to their specific property, can be used to realise any other algebras.

Definition 2.1 Let V be an $\mathbb{F}$-vector space. A pair (A, ι), consisting of an $\mathbb{F}$-algebra A and a linear map $\iota \colon V \to A$, is called *free algebra* on V if it has the following *universal property*: for any $\mathbb{F}$-algebra B and any linear map $f \colon V \to B$, there exists a unique algebra morphism $\tilde{f} \colon A \to B$ such that $\tilde{f}\iota = f$.

© The Author(s), under exclusive license to Springer Nature Switzerland AG 2023
R. Kashaev, *A Course on Hopf Algebras*, Universitext,
https://doi.org/10.1007/978-3-031-26306-4_2

Remark 2.1 If we denote by $\mathrm{Alg}(A, B)$ the set of all algebra morphisms from A to B, and by $\mathrm{L}(V, W)$ the set of all linear maps from V to W, then the universal property of a free algebra $(A, \imath)$ on V is equivalent to the bijectivity property of the map

$$\mathrm{Hom}(\imath, B)\colon \mathrm{Alg}(A, B) \to \mathrm{L}(V, B), \quad f \mapsto f\imath,$$

for any algebra B.

Exercise 2.1 Let $(A, \imath)$ and $(B, \jmath)$ be two free algebras over one and the same $\mathbb{F}$-vector space V. Show that there exists a unique algebra isomorphism $f\colon A \to B$ such that $\jmath = f\imath$.

Example 2.1 Let $\mathbb{F}[x]$ be the algebra of $\mathbb{F}$-polynomials of one indeterminate x and $\imath\colon \mathbb{F} \to \mathbb{F}[x]$, the linear map defined by $\imath 1 = x$. Then, the pair $(\mathbb{F}[x], \imath)$ is a free algebra over the one dimensional vector space $\mathbb{F}$. Indeed, for any $\mathbb{F}$-algebra B and any linear map $f\colon \mathbb{F} \to B$, we define an algebra morphism $\tilde{f}\colon \mathbb{F}[x] \to B$ by $\tilde{f}(p(x)) = p(f1)$. Then for any $a \in \mathbb{F}$, we have

$$fa = af1 = a\tilde{f}x = a\tilde{f}\imath 1 = \tilde{f}\imath a \quad \Rightarrow \quad f = \tilde{f}\imath. \tag{2.1}$$

Now, if $h\colon \mathbb{F}[x] \to B$ is any other algebra morphism with the property $f = h\imath$, then, for any $p(x) \in \mathbb{F}[x]$, we have

$$h(p(x)) = p(hx) = p(h\imath 1) = p(f1) = p(\tilde{f}\imath 1) = p(\tilde{f}x) = \tilde{f}(p(x)) \Rightarrow h = \tilde{f}. \tag{2.2}$$

2.1.1 The Tensor Algebra

For a given $\mathbb{F}$-vector space V, we define recursively the tensor powers of V as follows:

$$V^{\otimes 0} = \mathbb{F}, \quad V^{\otimes(m+1)} = V^{\otimes m} \otimes V, \quad \forall m \in \omega, \tag{2.3}$$

with natural identifications

$$V^{\otimes m} \otimes V^{\otimes n} \simeq V^{\otimes(m+n)} \quad \forall m, n \in \omega. \tag{2.4}$$

Consider the vector space

$$T(V) := \bigoplus_{m \in \omega} V^{\otimes m} \tag{2.5}$$

together with the canonical projections and inclusions

$$p_m \colon T(V) \to V^{\otimes m}, \quad i_m \colon V^{\otimes m} \to T(V), \quad m \in \omega, \tag{2.6}$$

with the properties

$$p_m i_n = \delta_{m,n} \, \mathrm{id}_{V^{\otimes n}}, \quad \sum_{m \in \omega} i_m p_m = \mathrm{id}_{T(V)} . \tag{2.7}$$

Proposition-Definition 2.1 *The $\mathbb{F}$-vector space $T(V)$ is an algebra, called* tensor algebra *of V, with the product*

$$\mu := \mu_{T(V)} = \sum_{m,n \in \omega} i_{m+n}(p_m \otimes p_n) \colon T(V) \otimes T(V) \to T(V) \tag{2.8}$$

and the unit

$$\eta := \eta_{T(V)} = i_0 \colon \mathbb{F} \to T(V). \tag{2.9}$$

Proof By using the (easily verifiable) equalities

$$p_m \mu = \sum_{s=0}^{m} p_{m-s} \otimes p_s \quad \forall m \in \omega, \tag{2.10}$$

we calculate

$$\mu(\mu \otimes \mathrm{id}_{T(V)}) = \sum_{m,n \in \omega} i_{m+n} \left((p_m \mu) \otimes p_n \right)$$

$$= \sum_{m,n \in \omega} \sum_{s=0}^{m} i_{m+n} \left(p_{m-s} \otimes p_s \otimes p_n \right) = \sum_{n,s \in \omega} \sum_{m \geq s} i_{m+n} \left(p_{m-s} \otimes p_s \otimes p_n \right)$$

$$= \sum_{m,n,s \in \omega} i_{m+n+s} \left(p_m \otimes p_s \otimes p_n \right) \tag{2.11}$$

and

$$\mu(\mathrm{id}_{T(V)} \otimes \mu) = \sum_{m,n\geq 0} i_{m+n} (p_m \otimes (p_n \mu))$$

$$= \sum_{m,n\in\omega} \sum_{s=0}^{n} i_{m+n} (p_m \otimes p_s \otimes p_{n-s}) = \sum_{m,s\in\omega} \sum_{n\geq s} i_{m+n} (p_m \otimes p_s \otimes p_{n-s})$$

$$= \sum_{m,n,s\in\omega} i_{m+n+s} (p_m \otimes p_s \otimes p_n) \qquad (2.12)$$

thus establishing the associativity of μ.

The equalities

$$p_m \eta = \delta_{m,0} \quad \forall m \in \omega \qquad (2.13)$$

imply the unitality:

$$\mu(\eta \otimes \mathrm{id}_{T(V)}) = \sum_{m,n\in\omega} i_{m+n} ((p_m \eta) \otimes p_n) = \sum_{n\in\omega} i_n p_n = \mathrm{id}_{T(V)}, \qquad (2.14)$$

and

$$\mu(\mathrm{id}_{T(V)} \otimes \eta) = \sum_{m,n\in\omega} i_{m+n} (p_m \otimes (p_n \eta)) = \sum_{m\in\omega} i_m p_m = \mathrm{id}_{T(V)}. \qquad (2.15)$$

$\square$

Exercise 2.2 Show that the m-th iterated product in $T(V)$ is of the form

$$\mu^{(m)} = \sum_{(n_1,\ldots,n_m)\in\omega^m} i_{n_1+\cdots+n_m} (p_{n_1} \otimes \cdots \otimes p_{n_m}). \qquad (2.16)$$

2.1.2   The Universal Property of the Tensor Algebra

Lemma 2.1 *For any $\mathbb{F}$-vector space V, the canonical embeddings $i_m \colon V^{\otimes m} \to T(V)$, $m \in \omega$, factorise through the iterated products of the tensor algebra $T(V)$ according to the formula*

$$i_m = \mu^{(m)} i_1^{\otimes m} \quad \forall m \in \omega. \qquad (2.17)$$

Proof First, we observe that

$$i_{k+l} = \mu (i_k \otimes i_l) \quad \forall k, l \in \omega. \tag{2.18}$$

Indeed, we have

$$\mu (i_k \otimes i_l) = \sum_{m,n\in\omega} i_{m+n} (p_m \otimes p_n)(i_k \otimes i_l)$$

$$= \sum_{m,n\in\omega} i_{m+n} ((p_m i_k) \otimes (p_n i_l)) = \sum_{m,n\in\omega} \delta_{m,k}\delta_{n,l} i_{m+n} \left(\mathrm{id}_{V^{\otimes k}} \otimes \mathrm{id}_{V^{\otimes l}}\right)$$

$$= i_{k+l} \, \mathrm{id}_{V^{\otimes(k+l)}} = i_{k+l}. \tag{2.19}$$

Further, we proceed by recurrence. For $m = 0$, formula (2.17) holds true. Assuming it holds true for $m = k \geq 0$, we have

$$\mu^{(k+1)} i_1^{\otimes(k+1)} = \mu \left(\mu^{(k)} \otimes \mathrm{id}_{T(V)}\right)\left(i_1^{\otimes k} \otimes i_1\right)$$

$$= \mu \left(\left(\mu^{(k)} i_1^{\otimes k}\right) \otimes i_1\right) = \mu (i_k \otimes i_1) = i_{k+1}. \tag{2.20}$$

Theorem 2.1 *For any $\mathbb{F}$-vector space V, the pair $(T(V), \iota_V := i_1)$ is a free algebra over V. More precisely, for any algebra B and any linear map $f : V \to B$, the map*

$$\tilde{f} := \sum_{m\in\omega} \mu_B^{(m)} f^{\otimes m} p_m : T(V) \to B \tag{2.21}$$

is the unique algebra morphism such that $\tilde{f}\iota_V = f$.

Proof (*Uniqueness*) Assuming the existence of $\tilde{f}$, we have

$$\tilde{f} = \tilde{f} \, \mathrm{id}_{T(V)} = \sum_{m\in\omega} \tilde{f} i_m p_m = \sum_{m\in\omega} \tilde{f} \mu_{T(V)}^{(m)} i_1^{\otimes m} p_m$$

$$= \sum_{m\in\omega} \mu_B^{(m)} \tilde{f}^{\otimes m} i_1^{\otimes m} p_m = \sum_{m\in\omega} \mu_B^{(m)} (\tilde{f} i_1)^{\otimes m} p_m = \sum_{m\in\omega} \mu_B^{(m)} f^{\otimes m} p_m \tag{2.22}$$

where, in the third equality, we used Lemma 2.1, in the forth equality, the assumption that $\tilde{f}$ is an algebra morphism, and, in the last equality, the factorisation assumption $\tilde{f}\iota_V = f$.

(Existence) Let us show that the linear map $\tilde{f}$ defined in (2.21) is indeed an algebra morphism. We have

$$\mu_B(\tilde{f} \otimes \tilde{f}) = \sum_{m,n \in \omega} \mu_B \left(\mu_B^{(m)} \otimes \mu_B^{(n)}\right) (f^{\otimes m} \otimes f^{\otimes n})(p_m \otimes p_n)$$

$$= \sum_{m,n \in \omega} \mu_B^{(m+n)} f^{\otimes(m+n)} (p_m \otimes p_n) = \sum_{s \in \omega} \sum_{0 \le n \le s} \mu_B^{(s)} f^{\otimes s}(p_{s-n} \otimes p_n)$$

$$= \sum_{s \in \omega} \mu_B^{(s)} f^{\otimes s} p_s \mu_{T(V)} = \tilde{f} \mu_{T(V)} \qquad (2.23)$$

and

$$\tilde{f} \eta_{T(V)} = \sum_{m \in \omega} \mu_B^{(m)} f^{\otimes m} p_m i_0 = \sum_{m \in \omega} \delta_{m,0} \mu_B^{(m)} f^{\otimes m} \, \mathrm{id}_{\mathbb{F}}$$

$$= \mu_B^{(0)} f^{\otimes 0} = \eta_B \, \mathrm{id}_{\mathbb{F}} = \eta_B. \qquad (2.24)$$

(Factorisation) We have

$$\tilde{f} i_1 = \sum_{m \in \omega} \mu_B^{(m)} f^{\otimes m} p_m i_1 = \sum_{m \in \omega} \delta_{m,1} \mu_B^{(m)} f^{\otimes m} \, \mathrm{id}_V = \mu_B^{(1)} f = \mathrm{id}_B \, f = f.$$

$$(2.25)$$

2.1.3 Presentations of Algebras

Definition 2.2 Let A be an $\mathbb{F}$-algebra. A *(two sided) ideal* of A is a vector subspace $J \subset A$ that is stable under the left and right multiplications by elements of A, i.e. $AJ \subset J \supset JA$.

Exercise 2.3 Show that the kernel of any algebra morphism $f : A \rightarrow B$ is a two sided ideal of A.

Exercise 2.4 Let A be an algebra and J a two sided ideal of A. Show that the quotient vector space A/J is an algebra with the product

$$\mu_{A/J}((x + J) \otimes (y + J)) = \mu_A(x \otimes y) + J, \quad \forall x, y \in A, \qquad (2.26)$$

and the unit

$$\eta_{A/J} 1 = \eta_A 1 + J, \qquad (2.27)$$

and the canonical projection from A to A/J is an algebra morphism.

Definition 2.3 Let A be an $\mathbb{F}$-algebra. A *presentation* of A is an expression of the form $\mathbb{F}\langle E \mid R \rangle$ where E is a set and R is a subset of the tensor algebra $T(\mathbb{F}[E])$ such that $A \simeq T(\mathbb{F}[E])/J$ where J is the (two-sided) ideal of $T(\mathbb{F}[E])$ generated by R. In particular, the *algebra freely generated by* E is the tensor algebra $T(\mathbb{F}[E])$, which is also denoted as $\mathbb{F}\langle E \rangle$, has the presentation with $R = \emptyset$,

$$\mathbb{F}\langle E \mid \emptyset \rangle = \mathbb{F}\langle E \mid \rangle.$$

Let V be an $\mathbb{F}$-vector space, A an $\mathbb{F}$-algebra and $f: V \to A$ a linear map such that $\tilde{f}: T(V) \to A$ is surjective, where $\tilde{f}$ is the algebra morphism induced by f through the universal property of the pair $(T(V), \iota_V)$. For example, one can take for V the underlying vector space of A and $f = \mathrm{id}_A$. Then, one has the algebra isomorphism $T(V)/\ker(\tilde{f}) \simeq A$, where $\ker(\tilde{f})$ is the kernel of $\tilde{f}$ which is a two-sided ideal of $T(V)$. In this case, one can take a presentation $\mathbb{F}\langle B \mid R \rangle$ of the algebra A where $B \subset V$ is a linear basis of V, and $R \subset T(V)$ is a generating set for the kernel $\ker(\tilde{f})$, i.e.

$$\ker(\tilde{f}) = T(V) R \, T(V).$$

Example 2.2 Let $q \in \mathbb{F}$. The presentation $\mathbb{F}\langle E \mid R \rangle$ with

$$E = \{a, b\}, \quad R = \{ab - qba\} \tag{2.28}$$

corresponds to the algebra with the underlying vector space $\mathbb{F}[B]$ freely generated by the set $B := \{b^m a^n \mid m, n \in \omega\}$, and the algebra structure is given by the multiplication

$$\mu(b^k a^l \otimes b^m a^n) = q^{lm} b^{k+m} a^{l+n} \tag{2.29}$$

and the unit element

$$\eta 1 = b^0 a^0. \tag{2.30}$$

2.2 Construction of Coalgebras

As coalgebras are categorially dual objects to algebras, their construction is closely related to the construction of algebras.

2.2.1　Dual Coalgebras

Let $A = (A, \mu, \eta)$ be a finite dimensional $\mathbb{F}$-algebra. Then, the dual space $A^* := L(A, \mathbb{F})$ is a coalgebra with the coproduct

$$\Delta := \mu^* : A^* \to (A \otimes A)^* \simeq A^* \otimes A^* \tag{2.31}$$

and the counit

$$\epsilon := \eta^* : A^* \to \mathbb{F}^* \simeq \mathbb{F}. \tag{2.32}$$

Exercise 2.5 Let A be the three dimensional $\mathbb{F}$-algebra of upper triangular 2-by-2 matrices. Show that the coalgebra A^* contains a linear basis $\{a, b, c\} \subset A^*$ with the coproduct

$$\Delta a = a \otimes a, \quad \Delta b = a \otimes b + b \otimes c, \quad \Delta c = c \otimes c. \tag{2.33}$$

Remark 2.2 Let $A = (A, \mu, \eta)$ be an infinite dimensional $\mathbb{F}$-algebra. In this case, as the inclusion $A^* \otimes A^* \subset (A \otimes A)^*$ is strict, one does not necessarily have the inclusion $\mu^*(A^*) \subset A^* \otimes A^*$, and thus the dual space A^* is not necessarily a coalgebra. Nonetheless, the vector subspace $A^o := (\mu^*)^{-1}(A^* \otimes A^*) \subset A^*$ happens to be a coalgebra with the coproduct $\Delta = \mu^o := \mu^*|_{A^o}$ and the counit $\epsilon = \eta^o := \eta^*|_{A^o}$. This coalgebra is called the *restricted or finite dual* of A (see Chap. 3 for details).

2.2.2　Quotient Coalgebras

Definition 2.4 Let $C = (C, \Delta, \epsilon)$ be a $\mathbb{F}$-coalgebra. A *(two-sided) coideal* of C is a vector subspace $J \subset C$ such that

$$\epsilon(J) = 0, \quad \Delta(J) \subset J \otimes C + C \otimes J.$$

Let $C = (C, \Delta, \epsilon)$ be an $\mathbb{F}$-coalgebra and $J \subset C$ a coideal. Then, the quotient vector space C/J is a coalgebra with the coproduct

$$\Delta_{C/J}(x + J) = \sum_{(x)} (x_{(1)} + J) \otimes (x_{(2)} + J), \quad \forall x \in C,$$

and the counit $\epsilon_{C/J}(x + J) = \epsilon(x)$.

Conversely, if $f : C \to D$ is a coalgebra morphism, then its kernel $\ker(f) = f^{-1}(0)$ is a coideal, and the image $f(C)$ is a sub-coalgebra of D isomorphic to the quotient coalgebra $C/\ker(f)$. In particular, if f is surjective, then D is isomorphic to $C/\ker(f)$.

2.2.3 Direct Sum Coalgebras

Proposition-Definition 2.2 *For any family of coalgebras $\mathcal{F}$, let*

$$V := \oplus_{C \in \mathcal{F}} C$$

be the direct sum of the underlying vector spaces with the canonical projections $p_C : V \to C$ and inclusions $i_C : C \to V$ satisfying the relations

$$p_C i_D = \mathrm{id}_C\, \delta_{C,D}, \qquad \sum_{C \in \mathcal{F}} i_C p_C = \mathrm{id}_V.$$

Then, the triple (V, Δ, ϵ), where

$$\Delta := \sum_{C \in \mathcal{F}} (i_C \otimes i_C) \Delta_C p_C : V \to V \otimes V, \quad \epsilon := \sum_{C \in \mathcal{F}} \epsilon_C p_C : V \to \mathbb{F},$$

is a coalgebra called the direct sum coalgebra *of the family $\mathcal{F}$.* □

Proof (Coassociativity) We have

$$(\Delta \otimes \mathrm{id}_V)\Delta = \sum_{C \in \mathcal{F}} (\Delta \otimes \mathrm{id}_V)(i_C \otimes i_C)\Delta_C p_C = \sum_{C \in \mathcal{F}} ((\Delta i_C) \otimes i_C)\Delta_C p_C$$

$$= \sum_{C \in \mathcal{F}} (((i_C \otimes i_C)\Delta_C) \otimes i_C)\Delta_C p_C = \sum_{C \in \mathcal{F}} (i_C \otimes i_C \otimes i_C)(\Delta_C \otimes \mathrm{id}_C)\Delta_C p_C$$

and

$$(\mathrm{id}_V \otimes \Delta)\Delta = \sum_{C \in \mathcal{F}} (\mathrm{id}_V \otimes \Delta)(i_C \otimes i_C)\Delta_C p_C = \sum_{C \in \mathcal{F}} (i_C \otimes (\Delta i_C))\Delta_C p_C$$

$$= \sum_{C \in \mathcal{F}} (i_C \otimes ((i_C \otimes i_C)\Delta_C))\Delta_C p_C = \sum_{C \in \mathcal{F}} (i_C \otimes i_C \otimes i_C)(\mathrm{id}_C \otimes \Delta_C)\Delta_C p_C.$$

We see that the two expressions above coincide due to the coassociativity of the coproducts Δ_C.

(Counitality) We have

$$(\epsilon \otimes id_V)\Delta = \sum_{C\in\mathcal{F}}(\epsilon \otimes id_V)(i_C \otimes i_C)\Delta_C p_C = \sum_{C\in\mathcal{F}}((\epsilon i_C) \otimes i_C)\Delta_C p_C$$

$$= \sum_{C\in\mathcal{F}}(\epsilon_C \otimes i_C)\Delta_C p_C = \sum_{C\in\mathcal{F}} i_C(\epsilon_C \otimes id_C)\Delta_C p_C$$

$$= \sum_{C\in\mathcal{F}} i_C\, id_C\, p_C = \sum_{C\in\mathcal{F}} i_C p_C = id_V\,.$$

and

$$(id_V \otimes \epsilon)\Delta = \sum_{C\in\mathcal{F}}(id_V \otimes \epsilon)(i_C \otimes i_C)\Delta_C p_C = \sum_{C\in\mathcal{F}}(i_C \otimes (\epsilon i_C))\Delta_C p_C$$

$$= \sum_{C\in\mathcal{F}}(i_C \otimes \epsilon_C)\Delta_C p_C = \sum_{C\in\mathcal{F}} i_C(id_C \otimes \epsilon_C)\Delta_C p_C$$

$$= \sum_{C\in\mathcal{F}} i_C\, id_C\, p_C = \sum_{C\in\mathcal{F}} i_C p_C = id_V\,.$$

2.3 Construction of Bialgebras

Let C be an $\mathbb{F}$-coalgebra and $T(C)$ the tensor algebra of C (viewed as a vector space) with the canonical linear inclusion $\iota_C : C \to T(C)$. By the universal property of the pair $(T(C), \iota_C)$, the linear maps $(\iota_C \otimes \iota_C)\Delta_C : C \to T(C) \otimes T(C)$ and $\epsilon_C : C \to \mathbb{F}$, where $T(C) \otimes T(C)$ and $\mathbb{F}$ are considered as algebras, determine uniquely defined algebra morphisms $\Delta_{T(C)} : T(C) \to T(C) \otimes T(C)$ and $\epsilon_{T(C)} : T(C) \to \mathbb{F}$ such that

$$(\iota_C \otimes \iota_C)\Delta_C = \Delta_{T(C)}\iota_C \quad \text{and} \quad \epsilon_C = \epsilon_{T(C)}\iota_C \qquad (2.34)$$

so that ι_C appears to be a coalgebra morphism provided $(T(C), \Delta_{T(C)}, \epsilon_{T(C)})$ is a coalgebra.

Proposition 2.1 *For any $\mathbb{F}$-coalgebra C, the triple $(T(C), \Delta_{T(C)}, \epsilon_{T(C)})$ is an $\mathbb{F}$-coalgebra so that, by taking into account the algebra structure, $T(C)$ is an $\mathbb{F}$-bialgebra and the canonical inclusion $\iota_C : C \to T(C)$ is a coalgebra morphism.*

Proof In order to check the coassociativity of $\Delta_{T(C)}$, by using the first equality of (2.34) and the coassociativity of Δ_C, we calculate

$$(\Delta_{T(C)} \otimes \mathrm{id}_{T(C)})\Delta_{T(C)}\iota_C = (\Delta_{T(C)} \otimes \mathrm{id}_{T(C)})(\iota_C \otimes \iota_C)\Delta_C$$

$$= ((\Delta_{T(C)}\iota_C) \otimes \iota_C)\Delta_C = (((\iota_C \otimes \iota_C)\Delta_C) \otimes \iota_C)\Delta_C$$

$$= (\iota_C \otimes \iota_C \otimes \iota_C)(\Delta_C \otimes \mathrm{id}_C)\Delta_C = (\iota_C \otimes \iota_C \otimes \iota_C)(\mathrm{id}_C \otimes \Delta_C)\Delta_C$$

$$= (\iota_C \otimes ((\iota_C \otimes \iota_C)\Delta_C))\Delta_C = (\iota_C \otimes (\Delta_{T(C)}\iota_C))\Delta_C$$

$$= (\mathrm{id}_{T(C)} \otimes \Delta_{T(C)})((\iota_C \otimes \iota_C)\Delta_{T(C)}) = (\mathrm{id}_{T(C)} \otimes \Delta_{T(C)})\Delta_{T(C)}\iota_C. \qquad (2.35)$$

Thus, by the uniqueness part of the universal property of $(T(C), \iota_C)$, we conclude that $\Delta_{T(C)}$ is coassociative.

In order to check the counitality of $\epsilon_{T(C)}$, by using both equalities of (2.34) and the counitality of ϵ_C, we calculate

$$(\epsilon_{T(C)} \otimes \mathrm{id}_{T(C)})\Delta_{T(C)}\iota_C = (\epsilon_{T(C)} \otimes \mathrm{id}_{T(C)})(\iota_C \otimes \iota_C)\Delta_C$$

$$= ((\epsilon_{T(C)}\iota_C) \otimes \iota_C)\Delta_C = (\epsilon_C \otimes \iota_C)\Delta_C = \iota_C(\epsilon_C \otimes \mathrm{id}_C)\Delta_C = \iota_C \qquad (2.36)$$

and likewise

$$(\mathrm{id}_{T(C)} \otimes \epsilon_{T(C)})\Delta_{T(C)}\iota_C = \iota_C. \qquad (2.37)$$

Thus, again, the uniqueness part of the universal property of $(T(C), \iota_C)$ implies the counitality of $\epsilon_{T(C)}$. Thus, the triple $(T(C), \Delta_{T(C)}, \epsilon_{T(C)})$ is a coalgebra.

The fact that the maps $\Delta_{T(C)}$ and $\epsilon_{T(C)}$, by construction, are algebra morphisms immediately implies that $T(C)$ is an $\mathbb{F}$-bialgebra.   □

Definition 2.5 Let C be an $\mathbb{F}$-coalgebra. A pair (B, ι) consisting of an $\mathbb{F}$-bialgebra B and a coalgebra morphism $\iota\colon C \to B$ is called *free bialgebra* on C if it satisfies the following *universal property*: for any $\mathbb{F}$-bialgebra D and any coalgebra morphism $f\colon C \to D$, there exists a unique bialgebra morphism $\tilde{f}\colon B \to D$ such that $\tilde{f}\iota = f$.

Remark 2.3 Similarly to the previous Remark 2.1, if we denote by $\mathrm{Bialg}(A, B)$ the set of all bialgebra morphisms from A to B, and by $\mathrm{Coalg}(C, D)$ the set of all coalgebra morphisms from C to D, then the universal property of a free bialgebra (A, ι) on C is equivalent to the bijectivity property of the map

$$\mathrm{Hom}(\iota, B)\colon \mathrm{Alg}(A, B) \to \mathrm{Coalg}(C, B), \quad f \mapsto f\iota,$$

for any bialgebra B.

Theorem 2.2 *For any $\mathbb{F}$-coalgebra C, the pair $(T(C), \iota_C)$ is a free bialgebra on C.*

Proof Let B be a $\mathbb{F}$-bialgebra and $f: C \to B$ a coalgebra morphism. Since any bialgebra is an algebra, by the universal property of the tensor algebra, there exists a unique algebra morphism $\tilde{f}: T(C) \to B$ such that $f = \tilde{f}\iota_C$. Thus, it suffices to show that $\tilde{f}$ is also a coalgebra morphism.

By using the fact that f and ι_C are coalgebra morphisms, we have

$$\Delta_B \tilde{f}\iota_C = \Delta_B f = (f \otimes f)\Delta_C = ((\tilde{f}\iota_C) \otimes (\tilde{f}\iota_C))\Delta_C$$
$$= (\tilde{f} \otimes \tilde{f})(\iota_C \otimes \iota_C)\Delta_C = (\tilde{f} \otimes \tilde{f})\Delta_{T(C)}\iota_C \qquad (2.38)$$

and

$$\epsilon_B \tilde{f}\iota_C = \epsilon_B f = \epsilon_C = \epsilon_{T(C)}\iota_C. \qquad (2.39)$$

Thus, by the uniqueness part of the universal property of $(T(C), \iota_C)$, we conclude that

$$\Delta_B \tilde{f} = (\tilde{f} \otimes \tilde{f})\Delta_{T(C)} \quad \text{and} \quad \epsilon_B \tilde{f} = \epsilon_{T(C)}, \qquad (2.40)$$

that is $\tilde{f}$ is a coalgebra morphism.   □

2.3.1 Presentations of Bialgebras

Definition 2.6 Let B be an $\mathbb{F}$-bialgebra. A vector subspace of B is called *bi-ideal* if it is simultaneously a two-sided ideal and a coideal.

Exercise 2.6 Let B be a bialgebra and $J \subset B$ a bi-ideal. Show that the quotient vector space B/J carries a unique bialgebra structure such that the canonical projection $\pi: B \to B/J$ is a (surjective) bialgebra morphism. Conversely, for any bialgebra morphism $f: B \to D$, its kernel $\ker(f) = f^{-1}(0)$ is a bi-ideal of B. In particular, if f is surjective, then it induces a bialgebra isomorphism $D \simeq B/\ker(f)$.

Any bialgebra B is a quotient of a free bialgebra on a sub-coalgebra of B. It suffices to choose a sub-coalgebra $C \subset B$ which generates B as an algebra. By the universal property of $(T(C), \iota_C)$, the inclusion map $g: C \to B$ induces a surjective bialgebra morphism $\tilde{g}: T(C) \to B$ so that B is isomorphic to the quotient bialgebra $T(C)/\ker(\tilde{g})$. This allows us to extend presentations of algebras to *presentations of bialgebras*:

$$B \simeq \mathbb{F}\langle E \mid R, D \rangle \qquad (2.41)$$

where $E \subset C$ is a generating set, $R \subset \ker(\tilde{g})$ is a generating set of $\ker(\tilde{g})$, that is

$$\ker(\tilde{g}) = T(C)RT(C), \tag{2.42}$$

and

$$D := \{(e, \Delta_C e) \mid e \in E\}. \tag{2.43}$$

encodes the coalgebra structure.

Remark 2.4 For an element $(e, \Delta_C e)$ of D in (2.43), one can also use a less formal writing $\Delta e = \Delta_C e$, where Δ in the left hand side is the coproduct of the bialgebra that corresponds to the presentation.

Example 2.3 The presentation $\mathbb{F}\langle E \mid R, D \rangle$ with

$$E = \{a, b\}, \quad R = \{ab - ba\}, \quad D = \{(a, a \otimes a), (b, a \otimes b + b \otimes 1)\}$$

corresponds to the polynomial bialgebra $\mathbb{F}[a, b]$ with a group-like a and the coproduct $\Delta b = a \otimes b + b \otimes 1$ so that the alternative less formal writing of the set D (see Remark 2.4) is as follows:

$$D = \{\Delta a = a \otimes a, \; \Delta b = a \otimes b + b \otimes 1\}.$$

In this example, the counit can be determined uniquely from the counitality relations with the result $\epsilon a = 1$ and $\epsilon b = 0$. □

2.4   Construction of Hopf Algebras

Hopf algebras are constructed on the basis of bialgebras. The theory of free Hopf algebras on coalgebras has been developed originally by M. Takeuchi in the work [40].

2.4.1   Free Hopf Algebras on Coalgebras

Definition 2.7 Let C be a coalgebra. A pair (H, ι) consisting of a Hopf algebra H and a coalgebra morphism $\iota : C \to H$ is called a *free Hopf algebra* on C if for any Hopf algebra A and any coalgebra morphism $f : C \to A$, there exists a unique Hopf algebra morphism $\tilde{f} : H \to A$ such that $f = \tilde{f}\iota$.

For any $\mathbb{F}$-coalgebra C, we associate the direct sum coalgebra $C_* := \oplus_{n\in\omega} C_n$ where $C_n = C$ if n is even and $C_n = C^{\mathrm{cop}}$ if n is odd. Consider the linear maps

$$\varsigma := \sum_{n\in\omega} i_{n+1} p_n : C_* \to C_*^{\mathrm{cop}} \quad \text{and} \quad \iota_{C_*}\varsigma : C_* \to T(C_*)^{\mathrm{op,cop}} \tag{2.44}$$

where $i_n : C \to C_*$ is the (vector space) inclusion of C as the n-th direct summand, $p_n : C_* \to C$ is the projection to the n-th direct summand, and $(T(C_*), \iota_{C_*})$ is the free bialgebra on C_*.

Lemma 2.2 *The maps defined in (2.44) are coalgebra morphisms.*

Proof Denoting $\Delta := \Delta_C$, $\epsilon := \epsilon_C$, $\Delta_* := \Delta_{C_*}$ and $\epsilon_* := \epsilon_{C_*}$, we have the following decomposition formulae

$$\Delta_* = \sum_{n\in\omega} (i_n \otimes i_n) \Delta_n p_n, \quad \Delta_*^{\mathrm{op}} = \sum_{n\in\omega} (i_n \otimes i_n) \Delta_n^{\mathrm{op}} p_n, \quad \epsilon_* = \sum_{n\in\omega} \epsilon p_n \tag{2.45}$$

where Δ_n is the coproduct of C_n, that is $\Delta_n = \Delta$ if n is even and $\Delta_n = \Delta^{\mathrm{op}}$ if n is odd. This implies the equalities:

$$\Delta_{n+1}^{\mathrm{op}} = \Delta_n, \quad \forall n \in \omega. \tag{2.46}$$

(The case of ς) We have to verify the following two equalities

$$(\varsigma \otimes \varsigma)\Delta_* = \Delta_*^{\mathrm{op}}\varsigma, \quad \epsilon_*\varsigma = \epsilon_*. \tag{2.47}$$

For the first one, using the substitutions from (2.44) and (2.45), we calculate

$$(\varsigma \otimes \varsigma)\Delta_* = \sum_{k,l,n\in\omega} (i_{k+1} p_k \otimes i_{l+1} p_l)(i_n \otimes i_n)\Delta_n p_n$$

$$= \sum_{k,l,n\in\omega} (i_{k+1} p_k i_n \otimes i_{l+1} p_l i_n)\Delta_n p_n = \sum_{n\in\omega} (i_{n+1} \otimes i_{n+1})\Delta_n p_n \tag{2.48}$$

and

$$\Delta_*^{\mathrm{op}}\varsigma = \sum_{n,k\in\omega} (i_k \otimes i_k)\Delta_k^{\mathrm{op}} p_k i_{n+1} p_n = \sum_{n\in\omega} (i_{n+1} \otimes i_{n+1})\Delta_{n+1}^{\mathrm{op}} p_n$$

$$= \sum_{n\in\omega} (i_{n+1} \otimes i_{n+1})\Delta_n p_n \tag{2.49}$$

where, in the last equality, we used (2.46).

For the second equality of (2.47), we have

$$\epsilon_* \varsigma = \sum_{m,n \in \omega} \epsilon p_m \bar{\imath} n_{+1} p_n = \sum_{n \in \omega} \epsilon p_n = \epsilon_* \qquad (2.50)$$

Thus proving that ς is a morphism of coalgebras.

The second map in (2.44) is a morphism of coalgebras as a composition of morphisms of coalgebras. □

By the universal property of the free bialgebra, there exists a unique bialgebra morphism $\zeta : T(C_*) \to T(C_*)^{\mathrm{op,cop}}$ such that $\imath_{C_*} \varsigma = \zeta \imath_{C_*}$.

Proposition 2.2 *For any coalgebra C, the (two-sided) ideal J of $T(C_*)$ generated by the elements*

$$ux := \sum_{(x)} (\zeta x_{(1)}) x_{(2)} - (\epsilon_{T(C_*)} x) \eta_{T(C^*)} 1,$$

$$vx := \sum_{(x)} x_{(1)} \zeta x_{(2)} - (\epsilon_{T(C_*)} x) \eta_{T(C^*)} 1, \quad \forall x \in \imath_{C_*}(C_*), \qquad (2.51)$$

is a bi-ideal stable under the map ζ so that the corresponding quotient space $H(C) := T(C_*)/J$ is a bialgebra and the canonical projection $\pi : T(C_*) \to H(C)$ is a bialgebra morphism. Furthermore, $H(C)$ is a Hopf algebra with the antipode S induced from $\pi \zeta$ through the universal property of the quotient space, that is through the equation $\pi \zeta = S\pi$.

Proof Let us denote $\epsilon := \epsilon_{T(C_*)}$, $\eta := \eta_{T(C_*)}$ and $\Delta := \Delta_{T(C_*)}$. We have to show the following three properties of J:

$$\epsilon(J) = 0, \qquad (2.52)$$

$$\Delta(J) \subset J \otimes C + C \otimes J \qquad (2.53)$$

and

$$\zeta(J) \subset J. \qquad (2.54)$$

Due to the fact that ϵ, Δ, and ζ are bialgebra morphisms, it suffices to verify these properties only for the generating elements ux and vx for $x \in \imath_{C_*}(C_*)$. We check them explicitly in the case of elements ux and leave the case of vx as an exercise.

Property (2.52):

$$\epsilon(ux + x) = \sum_{(x)} (\epsilon \zeta x_{(1)}) \epsilon x_{(2)} = \epsilon \zeta x = \epsilon x.$$

Property (2.53):

$$\Delta ux + (\epsilon x)\eta 1 \otimes \eta 1 = \sum_{(x)} (\Delta \zeta x_{(1)})\Delta x_{(2)} = \sum_{(x)} ((\zeta \otimes \zeta)(\Delta^{op} x_{(1)}))(x_{(2)} \otimes x_{(3)})$$

$$= \sum_{(x)} (\zeta x_{(2)})x_{(3)} \otimes (\zeta x_{(1)})x_{(4)} = \sum_{(x)} (\zeta x_{(2)(1)})x_{(2)(1)} \otimes (\zeta x_{(1)})x_{(3)}$$

$$= \sum_{(x)} ((\zeta x_{(2)(1)})x_{(2)(1)} - (\epsilon x_{(2)})\eta 1) \otimes (\zeta x_{(1)})x_{(3)} + \sum_{(x)} \eta 1 \otimes (\zeta x_{(1)})x_{(2)}$$

$$= \sum_{(x)} ux_{(2)} \otimes (\zeta x_{(1)})x_{(3)} + \sum_{(x)} \eta 1 \otimes ux + (\epsilon x)\eta 1 \otimes \eta 1.$$

Property (2.54):

$$\zeta ux + (\epsilon x)\eta 1 = \sum_{(x)} (\zeta x_{(2)})\zeta \zeta x_{(1)} = \sum_{(x)} (\zeta x)_{(1)} \zeta (\zeta x)_{(2)} = u\zeta x + (\epsilon \zeta x)\eta 1$$

$$= u\zeta x + (\epsilon x)\eta 1 = uy + (\epsilon x)\eta 1$$

where $y := \zeta x \in \iota_{C_*}(C_*)$ due to the equality $\iota_{C_*} \zeta = \zeta \iota_{C_*}$.

Finally, the quotient map S induced by $\pi \zeta$ through the equation $\pi \zeta = S\pi$ is the inverse of $id_{H(C)}$ in the convolution algebra $End(H(C))$ due to the definition of the bi-ideal J and the fact that $H(C)$, as an algebra, is generated by the coalgebra C_*.   □

Theorem 2.3 *For any coalgebra C, let $\iota_C \colon C \to H(C)$ be defined as the composition $\pi \iota_{C_*} i_0 = \pi \iota_{C_*}|_C$ where $\pi \colon T(C_*) \to H(C) = T(C_*)/J$ is the canonical projection from the free bialgebra $(T(C_*), \iota_{C_*})$ on C_*. Then, the pair $(H(C), \iota_C)$ is a free Hopf algebra on C.*

Proof The map ι_C is a coalgebra morphism as a composition of coalgebra morphisms.

Let $f \colon C \to A$ be a coalgebra morphism to a Hopf algebra A. Then, the map $g \colon C_* \to A$ defined by

$$g := \sum_{n \geq 0} S_A^n f p_n \tag{2.55}$$

extends f in the sense that $g i_0 = g|_C = f$, and it is a coalgebra morphism as a consequence of the properties of powers of the antipode S_A with respect to the coproduct Δ_A and the counit ϵ_A, namely

$$(S_A^{2n} \otimes S_A^{2n})\Delta_A = \Delta_A S_A^{2n}, \quad (S_A^{2n+1} \otimes S_A^{2n+1})\Delta_A = \Delta_A^{op} S_A^{2n+1}, \quad \forall n \in \omega. \tag{2.56}$$

and

$$\epsilon_A S_A^n = \epsilon_A, \quad \forall n \in \omega. \tag{2.57}$$

Indeed, we calculate

$$(g \otimes g)\Delta_{C_*} = \sum_{k,m,n\in\omega} (S_A^m f p_{mi_k} \otimes S_A^n f p_{ni_k})\Delta_{C_k} p_k = \sum_{k\in\omega}(S_A^k f \otimes S_A^k f)\Delta_{C_k} p_k$$

$$= \sum_{k\in\omega}(S_A^k \otimes S_A^k)(f \otimes f)\Delta_{C_k} p_k = \sum_{k\in\omega}(S_A^{2k} \otimes S_A^{2k})(f \otimes f)\Delta_{C_{2k}} p_{2k}$$

$$+ \sum_{k\in\omega}(S_A^{2k+1} \otimes S_A^{2k+1})(f \otimes f)\Delta_{C_{2k+1}} p_{2k+1} = \sum_{k\in\omega}(S_A^{2k} \otimes S_A^{2k})(f \otimes f)\Delta_C p_{2k}$$

$$+ \sum_{k\in\omega}(S_A^{2k+1} \otimes S_A^{2k+1})(f \otimes f)\Delta_C^{op} p_{2k+1} = \sum_{k\in\omega}(S_A^{2k} \otimes S_A^{2k})\Delta_A f p_{2k}$$

$$+ \sum_{k\in\omega}(S_A^{2k+1} \otimes S_A^{2k+1})\Delta_A^{op} f p_{2k+1} = \sum_{k\in\omega}\Delta_A S_A^{2k} f p_{2k} + \sum_{k\in\omega}\Delta_A S_A^{2k+1} f p_{2k+1}$$

$$= \sum_{n\in\omega}\Delta_A S_A^n f p_n = \Delta_A \sum_{n\in\omega}S_A^n f p_n = \Delta_A g$$

and

$$\epsilon_A g = \sum_{n\geq 0}\epsilon_A S_A^n f p_n = \sum_{n\geq 0}\epsilon_A f p_n = \sum_{n\geq 0}\epsilon_C p_n = \epsilon_{C_*}.$$

By the universal property of the pair $(T(C_*), \iota_{C_*})$, there exists a unique bialgebra morphism $\tilde{g}: T(C_*) \to A$ such that $g = \tilde{g}\iota_{C_*}$ which verifies the equality $S_A\tilde{g} = \tilde{g}\varsigma$. Indeed, by using the property

$$p_n\varsigma = \sum_{k\geq 0} p_n i_{k+1} p_k = \sum_{k\geq 0} \delta_{n,k+1} p_k = p_{n-1} \quad \forall n \geq 1,$$

and the equality $\iota_{C_*}\varsigma = \zeta\iota_{C_*}$, we have

$$S_A\tilde{g}\iota_{C_*} = S_A g = \sum_{n\geq 0} S_A^{n+1} f p_n = \sum_{n\geq 1} S_A^n f p_{n-1}$$

$$= \sum_{n\geq 1} S_A^n f p_n \varsigma = g\varsigma = \tilde{g}\iota_{C_*}\varsigma = \tilde{g}\zeta\iota_{C_*}.$$

The obtained equality and the uniqueness part in the universal property of the pair $(T(C_*^{cop}), \iota_{C_*^{cop}})$ imply that $S_A\tilde{g} = \tilde{g}\zeta$.

Now, for any $x \in \iota_{C_*}(C_*)$, we have

$$\tilde{g}ux + (\epsilon_A \tilde{g}x)\eta_A 1 = \tilde{g}\sum_{(x)}(\zeta x_{(1)})x_{(2)} = \sum_{(x)}(\tilde{g}\zeta x_{(1)})\tilde{g}x_{(2)} = \sum_{(x)}(S_A \tilde{g}x_{(1)})\tilde{g}x_{(2)}$$

$$= \sum_{(x)}(S_A(\tilde{g}x)_{(1)})(\tilde{g}x)_{(2)} = (\epsilon_A \tilde{g}x)\eta_A 1$$

and similarly for vx. We conclude that $\tilde{g}(J) = 0$. By the universal property of the quotient space, there exists a unique bialgebra morphism $\tilde{f}: H(C) \to A$ such that $\tilde{g} = \tilde{f}\pi$. Furthermore, the equality

$$S_A \tilde{f}\pi = S_A \tilde{g} = \tilde{g}\zeta = \tilde{f}\pi\zeta = \tilde{f}S_{H(C)}\pi$$

and the surjectivity of π imply that $S_A \tilde{f} = \tilde{f}S_{H(C)}$, that is $\tilde{f}$ is a Hopf algebra morphism. Finally, we verify that

$$\tilde{f}\iota_C = \tilde{f}\pi\iota_{C_*}i_0 = \tilde{g}\iota_{C_*}i_0 = gi_0 = f.$$

$\square$

2.4.2 Presentations of Hopf Algebras

Let H be a Hopf algebra and $J \subset H$ a bi-ideal stable under the action of the antipode, that is $S_H(J) \subset J$. Then, the quotient bialgebra H/J is a Hopf algebra, and the canonical projection $\pi: H \to H/J$ is a (surjective) Hopf algebra morphism.

Indeed, the universal property of the quotient space implies that there exists a unique linear map $\tilde{S}: H/J \to H/J$ such that $\tilde{S}\pi = \pi S_H$. For any $x \in H$, by using the fact that π is a bialgebra morphism, we calculate the product $\tilde{S} * \mathrm{id}_{H/J}$ in the convolution algebra $\mathrm{End}(H/J)$:

$$(\tilde{S} * \mathrm{id}_{H/J})\pi x = \sum_{(x)}(\tilde{S}(\pi x)_{(1)})(\pi x)_{(2)} = \sum_{(x)}(\tilde{S}\pi x_{(1)})\pi x_{(2)}$$

$$= \sum_{(x)}(\pi S_H x_{(1)})\pi x_{(2)} = \pi \sum_{(x)}(S_H x_{(1)})x_{(2)} = \pi \eta_H \epsilon_H x = \eta_{H/J}\epsilon_{H/J}\pi x$$

$$(2.58)$$

and similarly for the product $\mathrm{id}_{H/J} * \tilde{S}$. Thus, $\tilde{S}$ is the antipode in the bialgebra H/J.

Conversely, for any Hopf algebra morphism $f: H \to L$, its kernel $\ker(f) := f^{-1}(0)$ is a bi-ideal of H stable under the action of S_H. In particular, if f is surjective, then it induces a Hopf algebra isomorphism $L \simeq H/\ker(f)$. This allows us to extend presentations of bialgebras to presentations of Hopf algebras by taking quotients of Hopf algebras with respect to bi-ideals stable by the antipode. The fact that the antipode is always unique allows to use bialgebra presentations also for Hopf algebras implicitly assuming the existence of the antipode. This means that one and the same presentation can have different meanings in the case of a bialgebra and a Hopf algebra. For example, as all group-like elements in a Hopf algebra are invertible, all group-like generators of a Hopf algebra presentation are assumed to be invertible, even if in the presentation it is not explicitly indicated that they are invertible.

Example 2.4 The bialgebra presentation of Example 2.3 corresponds also to a Hopf algebra presentation where the group-like element a is necessarily invertible. In this way, we arrive to the polynomial algebra $\mathbb{F}[a^{\pm 1}, b]$ with group-like element a and the coproduct $\Delta b = a \otimes b + b \otimes 1$, where the antipode can be recovered just by solving the equations corresponding to the invertibility axiom.

Indeed, in the case of the group-like element, the invertibility axiom states that

$$\eta 1 = \eta \epsilon a = (Sa)a = aSa$$

implying that a is an invertible element with the inverse $a^{-1} = Sa$.

In the case of element b, the invertibility axiom states that

$$0 = \eta \epsilon b = (Sa)b + Sb = aSb + b$$

implying that $Sb = -a^{-1}b$. □

Chapter 3
The Restricted Dual of an Algebra

As we already have seen in the previous chapters, if C is a coalgebra, then the dual space $C^* = L(C, \mathbb{F})$ is an algebra with the convolution product

$$\mu_{C^*} = \Delta^*|_{C^* \otimes C^*}.$$

However, the categorial duality between algebras and coalgebras does not allow us to conclude that the dual space of an algebra is a coalgebra with respect to the dual structural maps. The reason is that for a vector space V, the inclusion $V^* \otimes V^* \subset (V \otimes V)^*$ is strict if V is infinite dimensional. This means that, the dual vector space A^* of an algebra is a coalgebra with respect to the dual structural maps only if $\mu^*(A^*) \subset A^* \otimes A^*$. This motivates the definition of the restricted dual of an algebra.

Definition 3.1 The *restricted (or finite) dual* A^o of an algebra A is the vector subspace of A^* given by the inverse image of the tensor square of the dual vector space A^* by the dual of the product of A, i.e.

$$A^o := (\mu^*)^{-1}(A^* \otimes A^*). \tag{3.1}$$

3.1 The Restricted Dual and Finite Dimensional Representations

In this section, elements of the restricted dual A^o are characterised in terms of finite dimensional representations of A and A^o is shown to be a coalgebra with respect to the dual structural maps, that is $\mu^*(A^o) \subset A^o \otimes A^o$.

When A is finite dimensional, one always has the equality $A^o = A^*$. When A is infinite dimensional, A^o is a subspace of A^* which can be both the whole space

© The Author(s), under exclusive license to Springer Nature Switzerland AG 2023
R. Kashaev, *A Course on Hopf Algebras*, Universitext,
https://doi.org/10.1007/978-3-031-26306-4_3

$A^o = A^*$ or the trivial subspace $A^o = 0$. The result of this section implies that $A^o = 0$ in the case when A does not admit any finite dimensional representations.

In order to characterise elements of A^o, we consider the matrix elements of finite dimensional representations of A.

To begin with, let $\rho: A \to \mathbb{F}$ be an algebra morphism which corresponds to a one-dimensional representation. This means that ρ is a linear form with a specific behaviour with respect to the algebra structure of A, namely

$$\langle \rho, xy \rangle = \langle \rho, x \rangle \langle \rho, y \rangle$$

for any $x, y \in A$, and $\langle \rho, 1 \rangle = 1$. Let us rewrite $\langle \rho, xy \rangle$ as follows:

$$\langle \rho, xy \rangle = \langle \rho, \mu(x \otimes y) \rangle = \langle \mu^* \rho, x \otimes y \rangle. \tag{3.2}$$

By writing also

$$\langle \rho, x \rangle \langle \rho, y \rangle = \langle \rho \otimes \rho, x \otimes y \rangle, \tag{3.3}$$

we see that $\mu^* \rho = \rho \otimes \rho$, which means that ρ, considered as a linear form on A, is contained in the restricted dual of A.

Assume now, more generally, that V is an n-dimensional (left) A-module, i.e. that we have an algebra morphism $\lambda: A \to \text{End}(V)$. Let us choose a linear basis $\{v_i\}_{i \in \underline{n}} \subset V$ with $\underline{n} = \{0, 1, \ldots, n-1\}$, and for any $x \in A$ and $i \in \underline{n}$, consider the vector $(\lambda x)v_i$. As any other vector in V, it is a linear combination of the basis vectors where the coefficients are linear functions of x:

$$(\lambda x)v_i = \sum_{j \in \underline{n}} v_j \langle \lambda_{j,i}, x \rangle, \tag{3.4}$$

where the elements $\lambda_{j,i} \in A^*$ are called *matrix coefficients* of the representation λ with respect to the basis $\{v_i\}_{i \in \underline{n}}$. Writing

$$(\lambda(xy))v_i = \sum_{j \in \underline{n}} v_j \langle \lambda_{j,i}, xy \rangle = \sum_{j \in \underline{n}} v_j \langle \mu^* \lambda_{j,i}, x \otimes y \rangle \tag{3.5}$$

and

$$(\lambda x)(\lambda y)v_i = \sum_{k \in \underline{n}} (\lambda x)v_k \langle \lambda_{k,i}, y \rangle = \sum_{j,k \in \underline{n}} v_j \langle \lambda_{j,k}, x \rangle \langle \lambda_{k,i}, y \rangle$$

$$= \sum_{j,k \in \underline{n}} v_j \langle \lambda_{j,k} \otimes \lambda_{k,i}, x \otimes y \rangle,$$

and using the equality $\lambda(ab) = (\lambda a)(\lambda b)$, we conclude that

$$\mu^* \lambda_{j,i} = \sum_{k \in \underline{n}} \lambda_{j,k} \otimes \lambda_{k,i}, \quad \forall i, j \in \underline{n}, \tag{3.6}$$

i.e. $\{\lambda_{j,i}\}_{i,j \in \underline{n}} \subset A^o$ and $\mu^*(\{\lambda_{j,i}\}_{i,j \in \underline{n}}) \subset A^o \otimes A^o$.

Remark 3.1 The matrix coefficients $\{\lambda_{j,i}\}_{i,j \in \underline{n}}$ generate a finite dimensional sub-coalgebra of A^o which is an isomorphic image of the matrix coalgebra from Example 1.13.

Theorem 3.1 *The restricted dual A^o of any algebra A is the linear span of the matrix coefficients of all finite dimensional representations of A.*

Proof Taking into account the preceding consideration, it suffices to show that, for any non zero element f of A^o, there exists a finite dimensional (left) A-module V_f such that f is a linear combination of the matrix coefficients of this representation (with respect to some basis).

The dual space A^* is a left A-module corresponding to the dual right multiplications $R_x^* \in \mathrm{End}(A^*)$, where $x \in A$ and $R_x \in \mathrm{End}(A)$ is defined by $R_x y = yx$. Indeed, for any $x, y, z \in A$ and $\alpha \in \mathbb{F}$, we verify the linearity

$$R_{x+\alpha y} z = z(x + \alpha y) = zx + \alpha zy = R_x z + \alpha R_y z = (R_x + \alpha R_y) z$$

$$\Rightarrow R_{x+\alpha y} = R_x + \alpha R_y \Rightarrow R_{x+\alpha y}^* = R_x^* + \alpha R_y^*$$

and it is easily checked that

$$R_x^* R_y^* = (R_y R_x)^* = R_{xy}^*, \quad R_1^* = (\mathrm{id}_A)^* = \mathrm{id}_{A^*}. \tag{3.7}$$

Let $V_f := R_A^* f \subset A^*$ be the orbit of f with respect to this action of A on A^*. The linear dependence of R_x^* on x implies that the set V_f is a vector subspace of A^*, and the map $\lambda \colon A \to \mathrm{End}(V_f)$ defined by $\lambda x = R_x^*|_{V_f}$ is an algebra morphism.

The condition $f \in A^o$ implies that

$$\mu^* f = \sum_{i \in \underline{n}} g_i \otimes h_i \tag{3.8}$$

for some $n \in \mathbb{Z}_{>0}$ and $g, h \in (A^*)^{\underline{n}}$. The calculation

$$\langle R_x^* f, y \rangle = \langle f, yx \rangle = \langle \mu^* f, y \otimes x \rangle = \sum_{i \in \underline{n}} \langle g_i, y \rangle \langle h_i, x \rangle = \left\langle \sum_{i \in \underline{n}} g_i \langle h_i, x \rangle, y \right\rangle \tag{3.9}$$

shows that for any $x \in A$, the element $R_x^* f$ finds itself in the linear span of the elements $\{g_i\}_{i \in \underline{n}}$:

$$R_x^* f = \sum_{i \in \underline{n}} g_i \langle h_i, x \rangle. \tag{3.10}$$

Thus, $m := \dim(V_f) \leq n < \infty$.

Let $\{v_i\}_{i \in \underline{m}}$ be a linear basis of V_f with $\underline{m} = \{0, 1, \ldots, m-1\}$. Then, for any $x \in A$, we have

$$R_x^* f = \sum_{i \in \underline{m}} v_i \langle w_i, x \rangle \tag{3.11}$$

for some $w \in (A^*)^{\underline{m}}$. In particular,

$$f = R_1^* f = \sum_{i \in \underline{m}} v_i \langle w_i, 1 \rangle. \tag{3.12}$$

Let $z \in A^{\underline{m}}$ be such that

$$v_i = R_{z_i}^* f, \quad \forall i \in \underline{m}. \tag{3.13}$$

We have

$$(\lambda x) v_i = (\lambda x) R_{z_i}^* f = R_{x z_i}^* f = \sum_{j \in \underline{m}} v_j \langle w_j, x z_i \rangle = \sum_{j \in \underline{m}} v_j \langle R_{z_i}^* w_j, x \rangle. \tag{3.14}$$

Thus, the matrix coefficients $\{\lambda_{i,j}\}_{i,j \in \underline{m}}$ of the representation λ, corresponding to the basis $\{v_i\}_{i \in \underline{m}}$, are given by

$$\lambda_{i,j} = R_{z_j}^* w_i, \quad \forall i, j \in \underline{m}. \tag{3.15}$$

Let us show that f is a linear combination of $\lambda_{i,j}$'s.

By using (3.11), for any $x \in A$, we write

$$\langle f, x \rangle = \langle R_x^* f, 1 \rangle = \sum_{i \in \underline{m}} \langle v_i, 1 \rangle \langle w_i, x \rangle = \left\langle \sum_{i \in \underline{m}} \langle v_i, 1 \rangle w_i, x \right\rangle \tag{3.16}$$

which means that

$$f = \sum_{i \in \underline{m}} \langle v_i, 1 \rangle w_i. \tag{3.17}$$

By applying $R^*_{z_j}$ to both sides of this decomposition, we obtain

$$v_j = R^*_{z_j} f = \sum_{i \in \underline{m}} \langle v_i, 1 \rangle R^*_{z_j} w_i = \sum_{i \in \underline{m}} \langle v_i, 1 \rangle \lambda_{i,j}. \qquad (3.18)$$

Finally, by substituting this into (3.12), we obtain

$$f = \sum_{i,j \in \underline{m}} \langle v_i, 1 \rangle \langle w_j, 1 \rangle \lambda_{i,j}. \qquad (3.19)$$

Corollary 3.1 *For any algebra A, one has the inclusion $\mu^*(A^o) \subset A^o \otimes A^o$.*

This follows immediately from (3.6).

Exercise 3.1 For any algebra A let $\iota_A: A^o \to A^*$ be the canonical inclusion map. Let $f: A \to B$ be an algebra morphism. Show that

1. there exists a unique coalgebra morphism $f^o: B^o \to A^o$ such that

$$f^* \iota_B = \iota_A f^o;$$

2. $(\mathrm{id}_A)^o = \mathrm{id}_{A^o}$;
3. $(fg)^o = g^o f^o$ for any algebra morphism $g: Z \to A$;

Remark 3.2 The parts (2) and (3) of Exercise 3.1 reflect the functorial nature of the restricted dual which directly follows from the functorial nature of the duality correspondence for vector spaces. The restricted dual is, in fact, a contravariant functor from the category **Alg**$_\mathbb{F}$ of $\mathbb{F}$-algebras to the category **Coalg**$_\mathbb{F}$ of $\mathbb{F}$-coalgebras. One can also show that there exists a natural equivalence

$$\mathrm{Hom}_{\mathbf{Alg}_\mathbb{F}}(A, C^*) \simeq \mathrm{Hom}_{\mathbf{Coalg}_\mathbb{F}}(C, A^o), \quad \forall (A, C) \in \mathbf{Alg}_\mathbb{F} \times \mathbf{Cog}_\mathbb{F}. \qquad (3.20)$$

Exercise 3.2 Let $f: A \to B$ be a surjective morphism of algebras. Show that $f^o: B^o \to A^o$ is an injective morphism of coalgebras.

3.1.1 An Algebra with Trivial Restricted Dual

Theorem 3.1 implies that, if an algebra A does not admit finite dimensional representations, then its restricted dual is trivial, i.e. $A^o = 0$. For example, consider the Heisenberg subalgebra A_{Heis} of $\mathrm{End}(\mathbb{C}[z])$ generated by the multiplication and differentiation operators x and ∂ defined by

$$x(p(z)) = zp(z), \quad \partial(p(z)) = \frac{\mathrm{d}p(z)}{\mathrm{d}z}, \quad \forall p(z) \in \mathbb{C}[z]. \qquad (3.21)$$

They satisfy the commutation relation

$$\partial x - x \partial = \mathrm{id}_{\mathbb{C}[z]}. \tag{3.22}$$

The Heisenberg algebra does not admit finite dimensional representations. Indeed, assume that there is an algebra homomorphism $\lambda \colon A_{\mathrm{Heis}} \to \mathrm{End}(V)$, where $n :=$ $\dim(V) \in \mathbb{Z}_{>0}$. By taking the trace of the identity

$$(\lambda \partial)(\lambda x) - (\lambda x)(\lambda \partial) = \mathrm{id}_V, \tag{3.23}$$

and using the cyclic property of the trace, we obtain the equality $0 = n > 0$ which is a contradiction. Thus, $(A_{\mathrm{Heis}})^o = 0$.

3.1.2   An Infinite Dimensional Algebra A with $A^o = A^*$

Let V be an infinite dimensional vector space. Define an algebra A_V which, as a vector space, is the direct sum $\mathbb{F} \oplus V$ and the product

$$\mu((\alpha, v) \otimes (\beta, w)) = (\alpha, v)(\beta, w) = (\alpha\beta, \alpha w + \beta v) \tag{3.24}$$

Let $p \in A_V^*$ be the linear form defined by

$$\langle p, (\alpha, v) \rangle = \alpha. \tag{3.25}$$

For any $f \in A_V^*$, we have

$$\langle \mu^* f, (\alpha, v) \otimes (\beta, w) \rangle = \langle f, (\alpha\beta, \alpha w + \beta v) \rangle$$
$$= \langle f, (1, 0) \rangle \alpha\beta + \langle f, (0, \alpha w + \beta v) \rangle = \langle f, (1, 0) \rangle \alpha\beta + \alpha \langle f, (0, w) \rangle + \beta \langle f, (0, v) \rangle$$
$$= -\langle f, (1, 0) \rangle \alpha\beta + \alpha \langle f, (\beta, w) \rangle + \beta \langle f, (\alpha, v) \rangle$$
$$= -\langle f, (1, 0) \rangle \langle p \otimes p, (\alpha, v) \otimes (\beta, w) \rangle$$
$$+ \langle p \otimes f, (\alpha, v) \otimes (\beta, w) \rangle + \langle f \otimes p, (\alpha, v) \otimes (\beta, w) \rangle$$
$$= \langle p \otimes f + f \otimes p - \langle f, (1, 0) \rangle p \otimes p, (\alpha, v) \otimes (\beta, w) \rangle. \tag{3.26}$$

Thus, $f \in A_V^o$ with

$$\mu^* f = p \otimes f + f \otimes p - \langle f, (1, 0) \rangle p \otimes p. \tag{3.27}$$

3.2 The Restricted Dual of the Tensor Product of Two Algebras

Lemma 3.1 *For any algebras A and B, the canonical embedding*

$$\alpha_{A,B} : A^o \otimes B^o \hookrightarrow (A \otimes B)^o \tag{3.28}$$

is a coalgebra isomorphism such that, for any pair of algebra morphisms $f : A \to U$ and $g : B \to V$, one has the equality

$$(f \otimes g)^o \alpha_{U,V} = \alpha_{A,B}(f^o \otimes g^o). \tag{3.29}$$

Proof

(1) Let A and B be algebras. Define the canonical algebra inclusions

$$\iota : A \hookrightarrow A \otimes B, \quad \jmath : B \hookrightarrow A \otimes B,$$
$$\iota x = x \otimes 1_B, \quad \jmath y = 1_A \otimes y, \quad \forall (x, y) \in A \times B. \tag{3.30}$$

Denoting $\alpha := \alpha_{A,B}$, let us show that the map

$$\beta := (\iota^o \otimes \jmath^o)\Delta_{(A \otimes B)^o} : (A \otimes B)^o \to A^o \otimes B^o \tag{3.31}$$

is the inverse of α.

For any $(\varphi, x, y) \in (A \otimes B)^o \times A \times B$, denoting $\Delta := \Delta_{(A \otimes B)^o}$, we have

$$\langle \alpha\beta\varphi, x \otimes y \rangle = \langle \beta\varphi, x \otimes y \rangle = \langle \Delta\varphi, \iota x \otimes \jmath y \rangle = \langle \varphi, (\iota x)(\jmath y) \rangle = \langle \varphi, x \otimes y \rangle \tag{3.32}$$

implying that β is a right inverse of α, and, for any $(f, g, x, y) \in A^o \times B^o \times A \times B$, we also have

$$\langle \beta\alpha(f \otimes g), x \otimes y \rangle = \langle \beta(f \otimes g), x \otimes y \rangle = \langle \Delta(f \otimes g), \iota x \otimes \jmath y \rangle$$
$$= \langle f \otimes g, (\iota x)(\jmath y) \rangle = \langle f \otimes g, x \otimes y \rangle \tag{3.33}$$

implying that β is a left inverse of α.

(2) In order to show that $\alpha_{A,B}$ is a morphism of coalgebras, it suffices to show that

$$\Delta_{(A \otimes B)^o}\alpha_{A,B} = (\alpha_{A,B} \otimes \alpha_{A,B})\Delta_{A^o \otimes B^o} \tag{3.34}$$

and

$$\epsilon_{(A \otimes B)^o} \alpha_{A,B} = \epsilon_{A^o} \otimes \epsilon_{B^o}. \tag{3.35}$$

Indeed, for any $(\varphi, \psi) \in A^o \times B^o$ and $(x, y, u, v) \in A^2 \times B^2$, we have

$$\langle \Delta_{(A \otimes B)^o} \alpha_{A,B} (\varphi \otimes \psi), x \otimes u \otimes y \otimes v \rangle = \langle \varphi \otimes \psi, xy \otimes uv \rangle = \langle \varphi, xy \rangle \langle \psi, uv \rangle$$

$$= \langle \Delta_{A^o} \varphi, x \otimes y \rangle \langle \Delta_{B^o} \psi, u \otimes v \rangle = \langle (\Delta_{A^o} \varphi) \otimes (\Delta_{B^o} \psi), x \otimes y \otimes u \otimes v \rangle$$

$$= \langle \Delta_{A^o \otimes B^o} (\varphi \otimes \psi), x \otimes u \otimes y \otimes v \rangle = \langle (\alpha_{A,B} \otimes \alpha_{A,B}) \Delta_{A^o \otimes B^o} (\varphi \otimes \psi), x \otimes u \otimes y \otimes v \rangle$$

and

$$\langle \epsilon_{(A \otimes B)^o} \alpha_{A,B}, \varphi \otimes \psi \rangle = \langle \varphi \otimes \psi, \eta_{A \otimes B} 1 \rangle$$

$$= \langle \varphi \otimes \psi, \eta_A 1 \otimes \eta_B 1 \rangle = \langle \varphi, \eta_A 1 \rangle \langle \psi, \eta_B 1 \rangle$$

$$= \langle \epsilon_{A^o}, \varphi \rangle \langle \epsilon_{B^o}, \psi \rangle = \langle \epsilon_{A^o} \otimes \epsilon_{B^o}, \varphi \otimes \psi \rangle.$$

(3) Let $f: A \rightarrow U$ and $g: B \rightarrow V$ be algebra morphisms. For any quadruple $(\varphi, \psi, x, y) \in U^o \times V^o \times A \times B$, we have

$$\langle (f \otimes g)^o \alpha_{U,V} (\varphi \otimes \psi), x \otimes y \rangle = \langle \varphi \otimes \psi, fx \otimes gy \rangle = \langle \varphi, fx \rangle \langle \psi, gy \rangle$$

$$= \langle f^o \varphi, x \rangle \langle g^o \psi, y \rangle = \langle f^o \varphi \otimes g^o \psi, x \otimes y \rangle = \langle \alpha_{A,B} (f^o \otimes g^o)(\varphi \otimes \psi), x \otimes y \rangle.$$

3.3   The Restricted Dual of a Hopf Algebra

The restricted dual H^o of a Hopf algebra H is defined as the restricted dual of the underlying algebra. In this subsection we show that the Hopf algebra operations of H imply that the restricted dual is itself a Hopf algebra.

Exercise 3.3 Let $f: X \rightarrow U$ and $g: Y \rightarrow V$ be two linear maps between vector spaces. Show that

$$(f \otimes g)^* |_{U^* \otimes V^*} = f^* |_{U^*} \otimes g^* |_{V^*}.$$

Proposition 3.1 *For any Hopf algebra $H = (H, \mu, \eta, \Delta, \epsilon, S)$, the restricted dual H^o is a Hopf algebra with respect to the dual structural maps*

$$\mu_{H^o} = \Delta^* |_{H^o \otimes H^o}, \quad \eta_{H^o} = \epsilon^* : 1 \mapsto \epsilon, \quad \Delta_{H^o} = \mu^* |_{H^o}, \quad \epsilon_{H^o} = \eta^o = \eta^* |_{H^o},$$

$$S_{H^o} = S^o = S^* |_{H^o}.$$

Proof By the functorial nature of the restricted dual, the vector space H^o is a coalgebra with the coproduct $\mu^*|_{H^o}$ and the counit η^o, and the algebra morphisms $\epsilon \colon H \to \mathbb{F}$ and $\Delta \colon H \to H \otimes H$ induce coalgebra morphisms $\epsilon^o \colon \mathbb{F} \to H^o$ and $\Delta^o \colon (H \otimes H)^o \to H^o$. By Lemma 3.1, the canonical inclusion

$$\alpha_{H,H} \colon H^o \otimes H^o \hookrightarrow (H \otimes H)^o$$

is an isomorphism of coalgebras and the composed map

$$\Delta^o \alpha_{H,H} \colon H^o \otimes H^o \to H^o$$

coincides with the restriction $\Delta^*|_{H^o \otimes H^o}$. This means that the triple

$$(H^o, \Delta^o \alpha_{H,H}, \epsilon^o)$$

is an algebra as a subalgebra of the convolution algebra H^*. Thus, the tuple

$$(H^o, \Delta^o \alpha_{H,H}, \epsilon^o, \mu^*|_{H^o}, \eta^o)$$

is a bialgebra.

Finally, we verify that S^o is the inverse of id_{H^o} in the convolution algebra $\mathrm{End}(H^o)$. By functoriality of the dual of a vector space, we have the equality

$$\epsilon^* \eta^* = \Delta^*(S \otimes \mathrm{id}_H)^* \mu^* \colon H^* \to H^* \tag{3.36}$$

which implies that

$$\eta_{H^o} \epsilon_{H^o} = \Delta^*(S \otimes \mathrm{id}_H)^* \mu^*|_{H^o} = \Delta^*(S \otimes \mathrm{id}_H)^*|_{H^o \otimes H^o} \Delta_{H^o}$$

$$= \Delta^*|_{H^o \otimes H^o}(S^o \otimes \mathrm{id}_{H^o}) \Delta_{H^o} = \mu_{H^o}(S^o \otimes \mathrm{id}_{H^o}) \Delta_{H^o}$$

where, in the third equality, we used Exercise 3.3. The second relation is verified similarly.   □

Chapter 4
The Restricted Dual of Hopf Algebras: Examples of Calculations

Let H be a Hopf algebra. In this chapter, we work out the structure of the Hopf algebra H^o in few simple examples. The fact that H^o a Hopf algebra (instead of being just a coalgebra) often facilitates its description.

It is useful to use the *(restricted) evaluation form*

$$E := E_H = \begin{pmatrix} E \\ \end{pmatrix} =: \begin{pmatrix} \\ \end{pmatrix} = \mathrm{ev}_H \,|_{H^o \otimes H} \in (H^o \otimes H)^*, \quad f \otimes x \mapsto \langle f, x \rangle$$

(4.1)

where we use solid line for H and dotted line for H^o. It has the following algebraic properties:

$$E(\mu_{H^o} \otimes \mathrm{id}_H) = E_{1,3} * E_{2,3} \iff \quad \begin{matrix} \end{matrix}$$

(4.2)

in the convolution algebra $(H^o \otimes H^o \otimes H)^*$,

$$E(\mathrm{id}_{H^o} \otimes \mu_H) = E_{1,2} * E_{1,3} \iff \quad \begin{matrix} \end{matrix}$$

(4.3)

in the convolution algebra $(H^o \otimes H \otimes H)^*$,

$$E(\eta_{H^o} \otimes \mathrm{id}_H) = \epsilon_H \iff \quad \begin{matrix} \end{matrix}$$

(4.4)

© The Author(s), under exclusive license to Springer Nature Switzerland AG 2023
R. Kashaev, *A Course on Hopf Algebras*, Universitext,
https://doi.org/10.1007/978-3-031-26306-4_4

and

$$E(\mathrm{id}_{H^o} \otimes \eta_H) = \epsilon_{H^o} \Leftrightarrow \quad = \quad .$$

(4.5)

The usefulness of the form E in elucidating the structure of H^o is the following *reproducing property*

$$E(\check{f} \otimes \mathrm{id}_H) = f, \quad \forall f \in H^o,$$

(4.6)

where the linear map $\check{f} \colon \mathbb{F} \to H^o$ is defined by $\check{f}1 = f$.

Let $\{e_i\}_{i \in I} \subset H$ be a linear basis and $\{e_i^*\}_{i \in I} \subset H^*$ the dual set of linear forms defined by the relations

$$\langle e_i^*, e_j \rangle = \delta_{i,j}, \quad \forall i, j \in I.$$

(4.7)

The evaluaton form E can be written as the following formal sum

$$E = \sum_{i \in I} e_i^{**} \otimes e_i^*$$

(4.8)

where $e_i^{**} \in (H^o)^*$ are defined by $\langle e_i^{**}, f \rangle = \langle f, e_i \rangle$. It has the property

$$\langle E, f \otimes x \rangle = \sum_{i \in I} \langle f, e_i \rangle \langle e_i^*, x \rangle.$$

(4.9)

By using the algebra structure of H^*, one can try to express the dual forms e_i^* in terms of some set of generating elements of H^*. Below, in concrete examples, we illustrate how this works in practice.

4.1   The Hopf Algebra $\mathbb{C}[x]$

The polynomial algebra $\mathbb{C}[x]$ is a Hopf algebra where x is a primitive element with the coproduct

$$\Delta x = x \otimes 1 + 1 \otimes x.$$

(4.10)

The monomials $\{x^n\}_{n \in \omega}$ form a linear basis, with the product

$$x^m x^n = x^{m+n}$$

(4.11)

and the coproduct

$$\Delta x^m = (\Delta x)^m = \sum_{k=0}^{m} \binom{m}{k} x^{m-k} \otimes x^k. \tag{4.12}$$

The counit and the antipode are easily calculated by using the axioms of the Hopf algebra.

For example, in order to calculate the counit applied to the generator x, we use the counitality property

$$(\epsilon \otimes \mathrm{id})\Delta x = x \Leftrightarrow \epsilon x + (\epsilon 1)x = x, \tag{4.13}$$

which, by using the equality $\epsilon 1 = 1$ valid in any Hopf algebra, implies that $\epsilon x = 0$.

For calculation of the image of the generator x under the action of the antipode, we use the invertibility axiom:

$$\mu(S \otimes \mathrm{id})\Delta x = \epsilon x \Leftrightarrow Sx + (S1)x = \epsilon x, \tag{4.14}$$

which, by the previous calculation and the equality $S1 = 1$ valid in any Hopf algebra, implies that $Sx = -x$.

Finally, by using the facts that the counit is an algebra morphism and the antipode is an algebra anti-morphism, we conclude that

$$\epsilon x^m = (\epsilon x)^m = \delta_{m,0}, \quad Sx^m = (Sx)^m = (-x)^m, \quad \forall m \in \omega. \tag{4.15}$$

Let us turn to the structure of the restricted dual of $\mathbb{C}[x]$.

One-dimensional representations of the algebra $\mathbb{C}[x]$ are indexed by the set of complex numbers $\mathbb{C}$. The matrix coefficient ρ_z corresponding to $z \in \mathbb{C}$ is defined as follows:

$$\langle \rho_z, x^n \rangle = z^n, \quad \forall n \in \omega. \tag{4.16}$$

It is a group-like element of the restricted dual $\mathbb{C}[x]^o$. The set of linear forms $\{\rho_z \mid z \in \mathbb{C}\}$ is closed under multiplication according to the multiplication rule

$$\rho_z \rho_w = \rho_{z+w} \quad \forall z, w \in \mathbb{C} \tag{4.17}$$

which is verified as follows:

$$\langle \rho_z \rho_w, x^n \rangle = \langle \rho_z \otimes \rho_w, \Delta x^n \rangle = \langle \rho_z \otimes \rho_w, (x \otimes 1 + 1 \otimes x)^n \rangle$$
$$= \left(\langle \rho_z, x \rangle \langle \rho_w, 1 \rangle + \langle \rho_z, 1 \rangle \langle \rho_w, x \rangle \right)^n = (z+w)^n = \langle \rho_{z+w}, x^n \rangle. \tag{4.18}$$

We define one more element of $\mathbb{C}[x]^o$, denoted ∂, by formally differentiating ρ_z with respect to z and then substituting $z = 0$:

$$\langle \partial, x^n \rangle = \left. \frac{\partial}{\partial z} \langle \rho_z, x^n \rangle \right|_{z=0} = \delta_{1,n}, \quad \forall n \in \omega. \tag{4.19}$$

Let us show that the linear form ∂ is a primitive element of the Hopf algebra $\mathbb{C}[x]^o$.
Indeed, for any $m, n \in \omega$, we have

$$\langle \Delta \partial, x^m \otimes x^n \rangle = \langle \partial, x^{m+n} \rangle = \delta_{1,m+n} = \delta_{0,m} \delta_{1,n} + \delta_{0,n} \delta_{1,m}$$

$$= \langle \epsilon, x^m \rangle \langle \partial, x^n \rangle + \langle \epsilon, x^n \rangle \langle \partial, x^m \rangle = \langle \epsilon \otimes \partial + \partial \otimes \epsilon, x^m \otimes x^n \rangle$$

$$\Rightarrow \Delta \partial = 1 \otimes \partial + \partial \otimes 1 \tag{4.20}$$

where we took into account the fact that the counit ϵ is the identity element in the restricted dual of a Hopf algebra.

Lemma 4.1 *The set of normalized monomials $\{\varphi_n := \partial^n / n!\}_{n \in \omega} \subset \mathbb{C}[x]^o$ forms the dual set of the monomial basis $\{x^n\}_{n \in \omega} \subset \mathbb{C}[x]$.*

Proof This can be proven by induction. Indeed, the case $n = 0$ holds true by the fact that $\varphi_0 = \epsilon$ (the counit of $\mathbb{C}[x]$) which is the identity element of the algebra $\mathbb{C}[x]^o$. Assuming that $\langle \varphi_{n-1}, x^m \rangle = \delta_{n-1,m}$ for a fixed $n \geq 1$ and arbitrary m, and the definition of ∂, we calculate

$$n \langle \varphi_n, x^m \rangle = \langle \varphi_{n-1} \partial, x^m \rangle = \langle \varphi_{n-1} \otimes \partial, \Delta x^m \rangle = \sum_{k=0}^{m} \binom{m}{k} \langle \varphi_{n-1} \otimes \partial, x^{m-k} \otimes x^k \rangle$$

$$= \sum_{k=0}^{m} \binom{m}{k} \delta_{n-1,m-k} \delta_{1,k} = \delta_{n,m} \sum_{k=0}^{n} \binom{n}{1} \delta_{1,k} = n \delta_{n,m}. \tag{4.21}$$

Remark 4.1 To keep track the evaluations of various linear forms on the linear basis $\{x^n\}_{n \in \omega}$, it is convenient to work with the exponential generating function for this basis

$$e^{\xi x} = \sum_{n=0}^{\infty} \frac{\xi^n}{n!} x^n. \tag{4.22}$$

For example, the set of evaluations $\langle \rho_z, x^n \rangle = z^n$, $n \in \omega$, can be collected into the evaluation on the generating function

$$\langle \rho_z, e^{\xi x} \rangle = \sum_{n=0}^{\infty} \frac{\xi^n}{n!} \langle \rho_z, x^n \rangle = \sum_{n=0}^{\infty} \frac{\xi^n}{n!} z^n = e^{\xi z}, \tag{4.23}$$

and similarly for the set of evaluations $\langle \partial^m, x^n \rangle = m! \delta_{m,n}$, $m, n \in \omega$,

$$\langle \partial^m, e^{\xi x} \rangle = \sum_{n=0}^{\infty} \frac{\xi^n}{n!} \langle \partial^m, x^n \rangle = \sum_{n=0}^{\infty} \frac{\xi^n}{n!} m! \delta_{m,n} = \xi^m. \tag{4.24}$$

Theorem 4.1 *The linear map*

$$v \colon \mathbb{C}[\mathbb{C}] \otimes \mathbb{C}[x] \to \mathbb{C}[x]^o, \quad \chi_z \otimes 1 \mapsto \rho_z, \ 1 \otimes x \mapsto \partial, \tag{4.25}$$

is an isomorphism of Hopf algebras.

Proof It is clear that the map v is a morphism of Hopf algebras. We only have to show that it is a bijection.

By Lemma 4.1, the dual evaluation form $E_{\mathbb{C}[x]}$ can be written as the following formal infinite sum

$$E_{\mathbb{C}[x]} = \sum_{n=0}^{\infty} \frac{1}{n!} (x^n)^{**} \otimes \partial^n \tag{4.26}$$

where we have used the convolution algebra structure of the space $\mathbb{C}[x]^*$. We will use this formula in order to show the surjectivity of the map v. To this end, we consider a representation of $\mathbb{C}[x]$ in a finite dimensional vector space with a fixed linear basis which corresponds to an algebra morphism

$$\lambda \colon \mathbb{C}[x] \to \mathrm{End}(\mathbb{C}^n). \tag{4.27}$$

Such a representation is completely characterised by the matrix $\lambda x = (\lambda_{i,j} x)_{i,j \in \underline{n}}$. By the Jordan block decomposition of λx, it suffices to assume that there exists a complex number z such that $A^n = 0$ where $A := \lambda x - z \, \mathrm{id}_{\mathbb{C}^n}$. Starting from the reproducing property (4.6) of the evaluation form, we proceed with the following calculation

$$\lambda_{i,j} = E_{\mathbb{C}[x]}(\check{\lambda}_{i,j} \otimes \mathrm{id}_{\mathbb{C}[x]}) = \sum_{k=0}^{\infty} \frac{1}{k!} \langle \lambda_{i,j}, x^k \rangle \partial^k = \sum_{k=0}^{\infty} \frac{1}{k!} \left((\lambda x)^k \right)_{i,j} \partial^k$$

$$= \sum_{k=0}^{\infty} \frac{1}{k!} \sum_{m=0}^{k} \binom{k}{m} (A^m)_{i,j} z^{k-m} \partial^k = \sum_{m=0}^{\infty} \sum_{k=m}^{\infty} \frac{(A^m)_{i,j} z^{k-m}}{(k-m)! m!} \partial^k$$

$$= \sum_{m,k=0}^{\infty} \frac{(A^m)_{i,j} z^k}{k! m!} \partial^{k+m} = \sum_{m=0}^{\infty} \frac{(A^m)_{i,j}}{m!} \partial^m \sum_{k=0}^{\infty} \frac{z^k}{k!} \partial^k = \sum_{m=0}^{n-1} \frac{(A^m)_{i,j}}{m!} \partial^m \rho_z$$

$$\tag{4.28}$$

where, in the last equality, we used the (formal) decomposition

$$\rho_z = E_{\mathbb{C}[x]}(\check{\rho}_z \otimes \mathrm{id}_{\mathbb{C}[x]}) = \sum_{k=0}^{\infty} \langle \rho_z, x^k \rangle \frac{\partial^k}{k!} = \sum_{k=0}^{\infty} z^k \frac{\partial^k}{k!}, \tag{4.29}$$

Thus, we obtain the formula

$$\lambda_{i,j} = v\left(\sum_{m=0}^{n-1} \frac{1}{m!} \left(A^m \right)_{i,j} \chi_z \otimes x^m \right) \tag{4.30}$$

implying surjectivity of the map v.

In order to show injectivity of v, assume that

$$\alpha := \sum_{w \in W} \chi_w \otimes p_w(x) \in \ker(v), \tag{4.31}$$

where $W \subset \mathbb{C}$ is a finite set of complex numbers and $p_w(x) \in \mathbb{C}[x]$ is a polynomial for each $w \in W$. This means that

$$v\alpha = \sum_{w \in W} \rho_w p_w(\partial) = 0. \tag{4.32}$$

By evaluating the latter on the generating series of the basis monomials $e^{\xi x}$, we obtain

$$0 = \langle v\alpha, e^{\xi x} \rangle = \sum_{w \in W} e^{w\xi} p_w(\xi). \tag{4.33}$$

The meaning of the obtained equality is that, by expanding the right hand side of (4.33) in power series in ξ, all the coefficients of that expansion vanish. On the other hand, as the series absolutely converges for any $\xi \in \mathbb{C}$, we can view the right hand side of (4.33) as an analytic function of ξ on the entire complex plane. It is impossible to identically vanish for this function unless all the polynomials $p_w(x)$ vanish. Indeed, let $s \in \mathbb{C}$ be such that the map $W \rightarrow \mathbb{C}_{\neq 0}$, $w \mapsto e^{sw}$, is injective. By taking the values $\xi = sn$, $n \in \omega$, in (4.33), we obtain the equalities

$$\sum_{w \in W} e^{wsn} p_w(sn) = 0, \quad \forall n \in \omega, \tag{4.34}$$

and, by Lemma 4.2 below, we conclude that $p_w(x) = 0$ for all $w \in W$, i.e. $\alpha = 0$ and thus the map v is injective. $\square$

Remark 4.2 The surjectivity of the map v in Theorem 4.1 can be proven by the following elegant and purely algebraic reasoning.[1] In formula (4.33), for each $w \in W$, the term $e^{w\xi} p_w(\xi)$ is a generalised eigenvector of the differentiation operator $\frac{\partial}{\partial \xi}$ corresponding to the eigenvalue w, that is there exists a positive integer $k \in \mathbb{Z}_{>0}$ such that $e^{w\xi} p_w(\xi)$ is in the kernel of the differential operator $(\frac{\partial}{\partial \xi} - w)^k$. This fact implies that the terms in the sum in (4.33) are linearly independent.

Lemma 4.2 *Let $\{p_w(x)\}_{w \in W} \subset \mathbb{C}[x]$ be a finite set of complex non-zero polynomials indexed by a (finite) set of non-zero complex numbers $W \subset \mathbb{C}_{\neq 0}$. Then there exists a non-negative integer $n \in \omega$ such that*

$$\sum_{w \in W} w^n p_w(n) \neq 0. \tag{4.35}$$

Proof The proof is by contradiction. Assume that

$$\sum_{w \in W} w^n p_w(n) = 0, \quad \forall n \in \omega. \tag{4.36}$$

Let $z \in \mathbb{C}$ be such that $|z| > |w|$ for all $w \in W$. Then, for any $w \in W$, the following series is absolutely convergent:

$$\sum_{n=0}^{\infty} w^n p_w(n) z^{-n} = p_w \left(-z \frac{\partial}{\partial z} \right) \sum_{n=0}^{\infty} \left(\frac{w}{z} \right)^n$$

$$= p_w \left(-z \frac{\partial}{\partial z} \right) \frac{z}{z - w} = q_w \left(\frac{z}{z - w} \right), \tag{4.37}$$

where

$$q_w(x) := p_w \left(x(x - 1) \frac{\partial}{\partial x} \right) x \tag{4.38}$$

is a polynomial of degree

$$\deg(q_w(x)) = 1 + \deg(p_w(x)) \geq 1 \tag{4.39}$$

which means that it is not a constant polynomial. On the other hand, summing (4.37) over all $w \in W$, we obtain the identity

$$\sum_{w \in W} q_w \left(\frac{z}{z - w} \right) = 0 \tag{4.40}$$

---

[1] This reasoning has been kindly suggested to the author by the anonymous referee.

which is impossible to satisfy, because due to the inequalities (4.39), z can approach arbitrarily close a pole w with the largest absolute value. $\square$

4.2   The Group Algebra $\mathbb{C}[\mathbb{Z}]$

The group algebra of the infinite cyclic group $\mathbb{Z}$ can be identified with the algebra of Laurent polynomials:

$$\mathbb{C}[\mathbb{Z}] \simeq \mathbb{C}[t, t^{-1}] \tag{4.41}$$

where the identification is given by $\chi_n \mapsto t^n$, so that the Hopf algebra structure in the space of Laurent polynomials is fixed by the condition that t is a group-like element. Let us determine the restricted dual of $\mathbb{C}[\mathbb{Z}]$.

Any one-dimensional representation of the algebra $\mathbb{C}[t, t^{-1}]$ is characterized by a non-zero complex number z corresponding to the image of the generator t. Denote the corresponding matrix coefficient by θ_z. It is a linear form on $\mathbb{C}[t, t^{-1}]$ defined by the formula

$$\langle \theta_z, t^n \rangle = z^n, \quad \forall n \in \mathbb{Z}, \tag{4.42}$$

which is a group-like element of the Hopf algebra $\mathbb{C}[t, t^{-1}]^o$. Moreover, we have the following multiplication rule

$$\theta_z \theta_w = \theta_{zw} \quad \forall z, w \in \mathbb{C}_{\neq 0} \tag{4.43}$$

which is checked as follows:

$$\langle \theta_z \theta_w, t^n \rangle = \langle \theta_z \otimes \theta_w, \Delta t^n \rangle = \langle \theta_z \otimes \theta_w, t^n \otimes t^n \rangle = z^n w^n = \langle \theta_{zw}, t^n \rangle. \tag{4.44}$$

In particular, θ_1 is nothing else than the unit element which we will denote by 1.

Differentiating formally θ_z with respect to z at $z = 1$, we define one more linear form ψ:

$$\langle \psi, t^n \rangle := \frac{\partial}{\partial z} \langle \theta_z, t^n \rangle \bigg|_{z=1} = n \quad \forall n \in \mathbb{Z} \tag{4.45}$$

which is a primitive element of $\mathbb{C}[t, t^{-1}]^o$ as shows the following calculation:

$$\langle \Delta \psi, t^m \otimes t^n \rangle = \langle \mu^* \psi, t^m \otimes t^n \rangle = \langle \psi, t^{m+n} \rangle = m + n = \langle \psi, t^m \rangle + \langle \psi, t^n \rangle$$
$$= \langle \psi, t^m \rangle \langle \epsilon, t^n \rangle + \langle \psi, t^n \rangle \langle \epsilon, t^m \rangle = \langle \psi \otimes \epsilon + \epsilon \otimes \psi, t^m \otimes t^n \rangle. \tag{4.46}$$

Exercise 4.1 Show that for any polynomial $p(x) \in \mathbb{C}[x]$, the following formula holds true:

$$\langle p(\psi), t^n \rangle = p(n) \quad \forall n \in \mathbb{Z}. \tag{4.47}$$

Remark 4.3 Pushing formula (4.47) even further, for any vector space valued function $g \colon \mathbb{Z} \to V$, we can associate a linear map $g(\psi) \colon \mathbb{C}[t, t^{-1}] \to V$ defined by

$$g(\psi)t^n = g(n), \quad \forall n \in \mathbb{Z}. \tag{4.48}$$

In particular, the linear forms θ_z can alternatively be denoted as z^ψ.

Let $\{\pi_n\}_{n \in \mathbb{Z}} \subset \mathbb{C}[t, t^{-1}]^*$ be the dual set of the canonical monomial basis $\{t^n\}_{n \in \mathbb{Z}} \subset \mathbb{C}[t, t^{-1}]$, with the convolution algebra structure

$$\pi_m \pi_n = \delta_{m,n} \pi_m, \quad \forall m, n \in \mathbb{Z}, \tag{4.49}$$

corresponding to the coalgebra structure of $\mathbb{C}[t, t^{-1}]$. Any linear form $f \in \mathbb{C}[t, t^{-1}]^*$ is decomposed into a formal sum

$$f = \sum_{n \in \mathbb{Z}} \langle f, t^n \rangle \pi_n \tag{4.50}$$

so that, in particular, we have

$$\psi = \sum_{n \in \mathbb{Z}} \langle \psi, t^n \rangle \pi_n = \sum_{n \in \mathbb{Z}} n \pi_n. \tag{4.51}$$

Theorem 4.2 *The linear map*

$$u \colon \mathbb{C}[C_{\neq 0}] \otimes \mathbb{C}[x] \to \mathbb{C}[t, t^{-1}]^o, \quad \chi_z \otimes 1 \mapsto \theta_z, \ 1 \otimes x \mapsto \psi, \tag{4.52}$$

is an isomorphism of Hopf algebras.

Proof The map u is evidently a Hopf algebra morphism. We start by showing that u is a surjective map.

Let $\lambda \colon \mathbb{C}[t, t^{-1}] \to \mathrm{End}(\mathbb{C}^n)$ be a finite dimensional representation which is completely characterized by the matrix $\lambda t = (\lambda_{i,j} t)_{i,j \in \underline{n}}$. By the Jordan block decomposition of λt, it suffices to assume that there exists a non-zero complex number z such that $(\lambda t - z \, \mathrm{id}_{\mathbb{C}^n})^n = 0$. Consider a set of elements $\{\varphi_{i,j}\}_{i,j \in \underline{n}}$ in

$\mathbb{C}[\mathbb{C}_{\neq 0}] \otimes \mathbb{C}[x]$ defined by the formula

$$\varphi_{i,j} := \chi_z \otimes \sum_{k=0}^{n-1} \binom{x}{k} \left(\left(z^{-1} \lambda t - \mathrm{id}_{\mathbb{C}^n} \right)^k \right)_{i,j}. \tag{4.53}$$

where

$$\binom{x}{k} := \frac{1}{k!} \prod_{j=0}^{k-1} (x - j) \in \mathbb{C}[x] \tag{4.54}$$

is the binomial polynomial of degree k. Let us show that $u\varphi_{i,j} = \lambda_{i,j}$.

For any $m \in \mathbb{Z}$, we have

$$\langle u\varphi_{i,j}, t^m \rangle = \left\langle \theta_z \sum_{k \in \underline{n}} \binom{\psi}{k}, t^m \right\rangle \left(\left(z^{-1} \lambda t - \mathrm{id}_{\mathbb{C}^n} \right)^k \right)_{i,j}$$

$$= z^m \sum_{k \in \underline{n}} \binom{m}{k} \left(\left(z^{-1} \lambda t - \mathrm{id}_{\mathbb{C}^n} \right)^k \right)_{i,j}$$

$$= z^m \sum_{k=0}^{\infty} \binom{m}{k} \left(\left(z^{-1} \lambda t - \mathrm{id}_{\mathbb{C}^n} \right)^k \right)_{i,j} = \left((\lambda t)^m \right)_{i,j} = \langle \lambda_{i,j}, t^m \rangle. \tag{4.55}$$

In order to show injectivity of u, assume that

$$\alpha := \sum_{w \in W} \chi_w \otimes p_w(x) \in \ker(u), \tag{4.56}$$

with a finite set of non-zero complex numbers $W \subset \mathbb{C}_{\neq 0}$ and some polynomials $p_w(x) \in \mathbb{C}[x]$. This means that

$$u\alpha = \sum_{w \in W} \theta_w p_w(\psi) = 0. \tag{4.57}$$

By evaluating the latter on the basis monomials, we obtain the equalities

$$0 = \langle u\alpha, t^n \rangle = \sum_{w \in W} w^n p_w(n), \quad \forall n \in \mathbb{Z}, \tag{4.58}$$

which, due to Lemma 4.2, imply that all the polynomials $p_w(x)$ identically vanish, i.e. $\alpha = 0$. □

4.3 The Hopf Algebra $J_\hbar$

Let us fix a scalar $\hbar \in \mathbb{C}$. We define a Hopf algebra with the presentation

$$J_\hbar = \mathbb{C}\langle a, b \mid ab - ba = \hbar a(a-1), \ \Delta b = a \otimes b + b \otimes 1, \ \Delta a = a \otimes a \rangle \quad (4.59)$$

where it is assumed that all grouplike elements are invertible.

Exercise 4.2 Show that for any $\hbar \in \mathbb{C}_{\neq 0}$, there exists an isomorphism of Hopf algebras $J_\hbar \simeq J_1$.

Exercise 4.3 Show that the set of monomials

$$\{b^m a^n \mid m \in \omega, \ n \in \mathbb{Z}\} \quad (4.60)$$

forms a linear basis in $J_\hbar$.

In the rest of this section, we provide a detailed description of the restricted dual of the Hopf algebra $J_\hbar$ assuming that $\hbar \neq 0$.

There exists a surjective morphism of Hopf algebras

$$J_\hbar \to \mathbb{C}[x], \quad a \mapsto 1, \ b \mapsto x, \quad (4.61)$$

which induces a Hopf algebra embedding (see Exercise 3.2)

$$\mathbb{C}[x]^o \simeq \mathbb{C}[\mathbb{C}] \otimes \mathbb{C}[x] \hookrightarrow J_\hbar^o \quad (4.62)$$

given by the linear forms $\{\rho_z\}_{z \in \mathbb{C}}$ and ∂ defined as follows:

$$\langle \rho_z, b^m a^n \rangle = z^m, \quad \langle \partial, b^m a^n \rangle = \delta_{m,1}, \quad \forall m \in \omega, \ \forall n \in \mathbb{Z}. \quad (4.63)$$

As in the case of $\mathbb{C}[x]^o$, it is convenient to rewrite the evaluations in (4.63) in terms of the generating series $e^{\xi b}$ for the powers of the generator b:

$$\langle \rho_z, e^{\xi b} a^n \rangle = e^{\xi z}, \quad \langle \partial, e^{\xi b} a^n \rangle = \xi, \quad \forall n \in \mathbb{Z}. \quad (4.64)$$

We also consider the following two-dimensional representation

$$\lambda \colon J_\hbar \to \mathrm{End}(\mathbb{C}^2), \quad a \mapsto \begin{pmatrix} 1 & 1 \\ 0 & 1 \end{pmatrix}, \quad b \mapsto \begin{pmatrix} 0 & 0 \\ 0 & \hbar \end{pmatrix}, \quad (4.65)$$

so that

$$\left(\langle \lambda_{i,j}, e^{\xi b} a^n \rangle \right)_{i,j \in \{1,2\}} = e^{\xi \lambda b} (\lambda a)^n = \begin{pmatrix} 1 & 0 \\ 0 & e^{\xi \hbar} \end{pmatrix} \begin{pmatrix} 1 & n \\ 0 & 1 \end{pmatrix} = \begin{pmatrix} 1 & n \\ 0 & e^{\xi \hbar} \end{pmatrix}. \quad (4.66)$$

We see that the lower left corner matrix coefficient is trivial, $\lambda_{2,1} = 0$, the diagonal matrix coefficients are contained in the Hopf sub-algebra $\mathbb{C}[x]^o$:

$$\lambda_{1,1} = \rho_0 = 1, \quad \lambda_{2,2} = \rho_\hbar, \tag{4.67}$$

while the upper right matrix coefficient gives a new linear form in $J_\hbar^o$:

$$\lambda_{1,2} =: \phi, \quad \langle \phi, e^{\xi b} a^n \rangle = n, \tag{4.68}$$

with the coproduct

$$\Delta\phi = 1 \otimes \phi + \phi \otimes \rho_\hbar. \tag{4.69}$$

Exercise 4.4 Let $\alpha, \beta, \gamma \in \mathbb{C}$ be such that $\alpha \neq \gamma$. Prove the following matrix equality:

$$\exp\left(\xi \begin{pmatrix} \alpha & \beta \\ 0 & \gamma \end{pmatrix}\right) = \begin{pmatrix} e^{\xi\alpha} & \beta(e^{\xi\alpha} - e^{\xi\gamma})/(\alpha - \gamma) \\ 0 & e^{\xi\gamma} \end{pmatrix}. \tag{4.70}$$

Lemma 4.3 *Let representation λ of $J_\hbar$ be defined in (4.65). For any $z \in \mathbb{C}$ and $n \in \mathbb{Z}$, the following generalization of (4.66) holds true:*

$$\lambda(e^{\xi(az+b)} a^n) = e^{\xi z} \begin{pmatrix} 1 & n + (e^{\xi\hbar} - 1)z/\hbar \\ 0 & e^{\xi\hbar} \end{pmatrix}. \tag{4.71}$$

Proof We have

$$e^{\xi(z\lambda a + \lambda b)} (\lambda a)^n = e^{\xi z} \exp\left(\xi \begin{pmatrix} 0 & z \\ 0 & \hbar \end{pmatrix}\right) (\lambda a)^n$$

$$= e^{\xi z} \begin{pmatrix} 1 & (e^{\xi\hbar} - 1)z/\hbar \\ 0 & e^{\xi\hbar} \end{pmatrix} \begin{pmatrix} 1 & n \\ 0 & 1 \end{pmatrix} = e^{\xi z} \begin{pmatrix} 1 & n + (e^{\xi\hbar} - 1)z/\hbar \\ 0 & e^{\xi\hbar} \end{pmatrix} \tag{4.72}$$

where we have used equality (4.70). $\qquad\qquad\qquad\qquad\qquad\qquad\qquad\square$

Proposition 4.1 *For any $z \in \mathbb{C}$, the linear forms ρ_z and ϕ satisfy the following relation*

$$\rho_z \phi \rho_{-z} = \phi + z(1 - \rho_\hbar)/\hbar. \tag{4.73}$$

Proof First, we calculate

$$\langle \rho_z \phi, e^{\xi b} a^n \rangle = \langle \rho_z \otimes \phi, \Delta(e^{\xi b} a^n) \rangle = \langle \rho_z \otimes \phi, e^{\xi(a \otimes b + b \otimes 1)}(a^n \otimes a^n) \rangle$$

$$= \langle \phi, e^{\xi(b+z)} a^n \rangle = e^{\xi z} \langle \lambda_{1,2}, e^{\xi b} a^n \rangle = e^{\xi z} n \qquad (4.74)$$

and similarly

$$\langle \phi \rho_z, e^{\xi b} a^n \rangle = \langle \phi \otimes \rho_z, \Delta(e^{\xi b} a^n) \rangle = \langle \phi \otimes \rho_z, e^{\xi(a \otimes b + b \otimes 1)}(a^n \otimes a^n) \rangle$$

$$= \langle \phi, e^{\xi(az+b)} a^n \rangle = \left[e^{\xi z \lambda a + \lambda b}(\lambda a)^n \right]_{1,2} = e^{\xi z}(n + (e^{\xi \hbar} - 1)z/\hbar) \qquad (4.75)$$

where we have used formula (4.71). Substructing this from (4.74), we obtain

$$\langle \rho_z \phi - \phi \rho_z, e^{\xi b} a^n \rangle = e^{\xi z}(1 - e^{\xi \hbar})z/\hbar = \langle \rho_z(1 - \rho_\hbar)z/\hbar, e^{\xi b} a^n \rangle \qquad (4.76)$$

which is equivalent to equality (4.73). $\qquad\qquad\square$

Corollary 4.1 *There exists a morphism of Hopf algebras*

$$J_\hbar \to J_\hbar^o, \quad a \mapsto \rho_{-\hbar}, \; b \mapsto -\hbar \phi \rho_{-\hbar}. \qquad (4.77)$$

Exercise 4.5 Show that

$$\partial \phi - \phi \partial = (1 - \rho_\hbar)/\hbar. \qquad (4.78)$$

Proposition 4.2 *For any $k \in \omega$, one has*

$$\langle \phi^k, e^{\xi b} a^n \rangle = n^k, \quad \forall n \in \mathbb{Z}, \qquad (4.79)$$

with the convention $0^0 = 1$.

Proof We proceed by induction on k. Equality (4.79) is true for $k = 0$. Assume that it is true for some $k \geq 0$. Then, by using (4.71) and (4.68), we have

$$\langle \phi^{k+1}, e^{\xi b} a^n \rangle = \langle \phi \otimes \phi^k, e^{\xi(a \otimes b + b \otimes 1)}(a^n \otimes a^n) \rangle$$

$$= \langle \phi^k, \left[e^{\xi(b\lambda a + \lambda b)}(\lambda a)^n \right]_{1,2} a^n \rangle = \langle \phi^k, e^{\xi b}(n + (e^{\xi \hbar} - 1)b/\hbar)a^n \rangle$$

$$= \left(n + (e^{\xi \hbar} - 1)\frac{1}{\hbar}\frac{\partial}{\partial \xi} \right) \langle \phi^k, e^{\xi b} a^n \rangle = \left(n + (e^{\xi \hbar} - 1)\frac{1}{\hbar}\frac{\partial}{\partial \xi} \right) n^k = n^{k+1}.$$

$$(4.80)$$

Exercise 4.6 Let X and Y be two linear transformations of a finite dimensional vector space such that

$$XY - YX = Y. \tag{4.81}$$

Show that

$$e^{zX} Y e^{-zX} = e^z Y, \quad \forall z \in \mathbb{C}. \tag{4.82}$$

Lemma 4.4 *For any finite dimensional representation* $\lambda \colon J_\hbar \to \text{End}(V)$, *the element* $\lambda a - \text{id}_V$ *is nilpotent.*

Proof The element $c := a^{-1} - 1 \in J_1$ is such that

$$bc - cb = -\hbar c. \tag{4.83}$$

By Exercise 4.6, we have the relation

$$e^{-z\lambda b} (\lambda c) e^{z\lambda b} = e^{\hbar z} \lambda c, \quad \forall z \in \mathbb{C}. \tag{4.84}$$

Let $\alpha \in \mathbb{C}$ be an eigenvalue of λc, and $v \in V \setminus \{0\}$ the corresponding eigenvector, i.e.

$$(\lambda c) v = \alpha v. \tag{4.85}$$

To any $z \in \mathbb{C}$, we associate the non zero vector

$$v_z := e^{z\lambda b} v. \tag{4.86}$$

Then, equality (4.84) implies that v_z is an eigenvector of λc corresponding to the eigenvalue $e^{\hbar z} \alpha$:

$$(\lambda c) v_z = (\lambda c) e^{z\lambda b} v = e^{z\lambda b} e^{\hbar z} (\lambda c) v = e^{z\lambda b} e^{\hbar z} \alpha v = e^{\hbar z} \alpha v_z, \quad \forall z \in \mathbb{C}. \tag{4.87}$$

This means that $\alpha = 0$, otherwise the set of eigenvalues of λc would be (uncountably) infinite and this not possible in a finite dimensional vector space. Thus, $\lambda c = (\text{id}_V - \lambda a)(\lambda a)^{-1}$ is a nilpotent element of $\text{End}(V)$. $\qquad\square$

The following theorem summarises the Hopf algebra structure of the restricted dual of $J_\hbar$.

Theorem 4.3 *Let* $\hbar \in \mathbb{C}_{\neq 0}$. *The Hopf algebra* $J_\hbar^\circ$ *is generated by the elements* $\{\rho_z\}_{z \in \mathbb{C}}$, ∂ *and* ϕ *defined by* (4.64) *and* (4.68) *with the coproducts*

$$\Delta \rho_z = \rho_z \otimes \rho_z, \quad \Delta \partial = 1 \otimes \partial + \partial \otimes 1, \quad \Delta \phi = 1 \otimes \phi + \phi \otimes \rho_\hbar \tag{4.88}$$

and satisfying the relations

$$\rho_z \rho_w = \rho_{z+w}, \qquad \rho_z \partial = \partial \rho_z, \tag{4.89}$$

$$\rho_z \phi \rho_{-z} = \phi + z(1 - \rho_\hbar)/\hbar, \qquad \partial \phi - \phi \partial = (1 - \rho_\hbar)/\hbar. \tag{4.90}$$

The set

$$B := \{\rho_z \partial^n \phi^m \mid z \in \mathbb{C}, \ m, n \in \omega\} \tag{4.91}$$

forms a linear basis of $J_\hbar^o$.

Proof The coproducts and the relations between the generating elements are already derived, see Theorem 4.1, Eqs. (4.65) and (4.69), Proposition 4.1, and Exercise 4.5.

Let

$$\lambda : J_\hbar \to \mathrm{End}(\mathbb{C}^n) \tag{4.92}$$

be arbitrary finite dimensional representation. By the formal formula for the dual evaluation form

$$\mathrm{ev}_{J_\hbar} = e^{b \otimes \partial} (a \otimes 1)^{1 \otimes \phi}, \tag{4.93}$$

we write for the matrix coefficients

$$\lambda_{i,j} = \mathrm{ev}_{J_\hbar}(\check{\lambda}_{ij} \otimes \mathrm{id}) = \left(e^{(\lambda b)\partial}(\lambda a)^\phi\right)_{i,j}. \tag{4.94}$$

By Lemma 4.4, there exists $L \in \omega$ such that $(\lambda a - \mathrm{id}_{\mathbb{C}^n})^L = 0$ so that

$$(\lambda a)^\phi = (\mathrm{id}_{\mathbb{C}^n} + \lambda a - \mathrm{id}_{\mathbb{C}^n})^\phi = \sum_{s=0}^{L-1} \binom{\phi}{s} (\lambda a - \mathrm{id}_{\mathbb{C}^n})^s, \tag{4.95}$$

thus obtaining a matrix valued polynomial in ϕ, i.e.

$$(\lambda a)^\phi \in \mathrm{End}(\mathbb{C}^n)[\phi]. \tag{4.96}$$

Let

$$S := \mathrm{Spec}(\lambda b) \subset \mathbb{C} \tag{4.97}$$

be the spectrum of the matrix λb. By Jordan's block structure of λb, for each eigenvalue $z \in S$, there exist a projection matrix $P_z \in \mathrm{End}(\mathbb{C}^n)$ and a positive integer $n_z \in \mathbb{Z}_{>0}$ such that

$$P_z \lambda b = (\lambda b) P_z, \quad P_z^2 = P_z, \quad P_z (\lambda b - z \, \mathrm{id}_{\mathbb{C}^n})^{n_z} = 0, \tag{4.98}$$

and also

$$P_z P_w = 0 \text{ if } z \neq w, \quad \sum_{z \in B} P_z = 1, \tag{4.99}$$

see Exercise 4.8 below. Thus, we can write

$$e^{(\lambda b)\partial} = \sum_{z \in S} \rho_z P_z e^{(\lambda b - z \, \mathrm{id}_{\mathbb{C}^n})\partial} = \sum_{z \in S} \rho_z P_z \sum_{r=0}^{n_z - 1} \frac{(\lambda b - z \, \mathrm{id}_{\mathbb{C}^n})^r}{r!} \partial^r = \sum_{z \in S} \rho_z P_z(\partial)$$

with matrix valued polynomials $p_z(x) \in \mathrm{End}(\mathbb{C}^n)[x]$. Putting everything together, we obtain a decomposition of the matrix coefficients into finite linear combinations of the set B:

$$\lambda_{i,j} = \sum_{z \in S} \rho_z \left(p_z(\partial)(\lambda a)^\phi \right)_{i,j}. \tag{4.100}$$

Let us prove now the linear independence of B. Assume that there exist finite sets of complex numbers $M_i \subset \mathbb{C}$ indexed by a finite set of non-negative integers $I \subset \omega$ such that

$$\sum_{i \in I} f_i(\partial)\phi^i = 0, \tag{4.101}$$

where

$$f_i(\partial) = \sum_{z \in M_i} \rho_z p_{i,z}(\partial), \quad p_{i,z}(x) \in \mathbb{C}[x]. \tag{4.102}$$

Evaluating at the basis elements $e^{\xi b} a^n$, we have

$$\sum_{i \in I} f_i(\xi) n^i = 0, \quad \forall n \in \mathbb{Z}, \tag{4.103}$$

and replacing ξ in this generating series by arbitrary complex numbers, we obtain

$$\sum_{i \in I} f_i(z) n^i = 0, \quad \forall n \in \mathbb{Z}, \forall z \in \mathbb{C}. \tag{4.104}$$

By applying Lemma 4.2 for each fixed z (with one polynomial $\sum_{i \in I} f_i(z) x^i$), we arrive at the equalities

$$f_i(z) := \sum_{w \in M_i} e^{wz} p_{i,w}(z) = 0, \quad \forall i \in I, \ \forall z \in \mathbb{C}, \tag{4.105}$$

which, again by Lemma 4.2 (with specifications $z = n \in \omega$), imply that all the polynomials $p_{i,w}(x)$ identically vanish. $\qquad \Box$

Exercise 4.7 Prove formula (4.100) by evaluating its both sides on the basis elements of $J_\hbar$.

Exercise 4.8 Let $T \in \operatorname{End}(V)$ with $\dim(V) < \infty$. By using Jordan's block decomposition of T show that there exists a set of positive integers $\{n_z\}_{z \in S} \subset \mathbb{Z}_{>0}$ indexed by a finite set of complex numbers $S \subset \mathbb{C}$ such that the polynomials $p_z(x) \in \mathbb{C}[x]$ defined by

$$p_z(x) := 1 - \left(1 - \prod_{w \in S \setminus \{z\}} \left(\frac{x-w}{z-w}\right)^{n_w}\right)^{n_z}, \quad \forall z \in S, \tag{4.106}$$

satisfy the relations

$$p_z(T) p_w(T) = \delta_{z,w} p_z(T), \quad \sum_{z \in S} p_z(T) = \operatorname{id}_V, \quad p_z(T)(T - z \operatorname{id}_V)^{n_z} = 0. \tag{4.107}$$

4.4 The Quantum Group B_q

Let q be a non-zero complex parameter. We define a Hopf algebra with the presentation

$$B_q = \mathbb{C}\langle a, b \mid ab = qba, \ \Delta a = a \otimes a, \ \Delta b = a \otimes b + b \otimes 1 \rangle \tag{4.108}$$

where, as in the case of the Hopf algebra $J_\hbar$, the generator a is grouplike and thus invertible. The Hopf algebra B_q is the smallest example of a *quantum group*, the term introduced by Drinfel'd in [11]. It is a q-deformation of $B_1 = J_0$ which is the algebra of regular functions $\mathbb{C}[\operatorname{Aff}_1(\mathbb{C})]$ on the affine linear algebraic group

$$\operatorname{Aff}_1(\mathbb{C}) := \mathbb{G}_a(\mathbb{C}) \rtimes \mathbb{G}_m(\mathbb{C}) \tag{4.109}$$

of invertible upper triangular complex 2-by-2 matrices of the form $\left(\begin{smallmatrix} a & b \\ 0 & 1 \end{smallmatrix}\right)$ where the Hopf algebra structure is canonically induced by the group structure of $\operatorname{Aff}_1(\mathbb{C})$, see [44].

In the algebra B_q, the monomials $b^m a^n$ with $m \in \omega$ and $n \in \mathbb{Z}$ form a closed under multiplication set with the following product

$$b^m a^n b^k a^l = q^{nk} b^{m+n} a^{n+l}. \tag{4.110}$$

Lemma 4.5 *The monomials $b^m a^n$ with $m \in \omega$ and $n \in \mathbb{Z}$ form a linear basis of B_q.*

Proof Consider the following action of the generators a and b in the space of polynomials $\mathbb{C}[x, y, y^{-1}]$:

$$a(f(x, y)) = yf(xq, y), \quad b(f(x, y)) = xf(x, y). \tag{4.111}$$

It is easily verified that this action gives rise to an algebra homomorphism from B_q to $\mathrm{End}(\mathbb{C}[x, y, y^{-1}])$:

$$a(b(f(x, y))) = a(xf(x, y)) = xqyf(xq, y) = qy(b(f(xq, y)))$$
$$= q(b(a(f(x, y)))). \tag{4.112}$$

Assume that there exists a finite set $\{c_{m,n}\}_{(m,n)\in I} \subset \mathbb{C}$ with $I \subset \omega \times \mathbb{Z}$ such that

$$\sum_{(m,n)\in I} c_{m,n} b^m a^n = 0. \tag{4.113}$$

By using the action (4.111), we have

$$\sum_{(m,n)\in I} c_{m,n} x^m y^n f(xq^n, y) = 0, \quad \forall f(x, y) \in \mathbb{C}[x, y, y^{-1}]. \tag{4.114}$$

Choosing $f(x, y) = 1$, we conclude that $c_{m,n} = 0$ for all $(m, n) \in I$. $\qquad\square$

In order to calculate the coproduct, we need the following *q-binomial formula*

$$(u + v)^m = \sum_{k=0}^m \begin{bmatrix} m \\ k \end{bmatrix}_q v^{m-k} u^k \tag{4.115}$$

where u and v are elements of any algebra which satisfy the relation $uv = qvu$, and

$$\begin{bmatrix} m \\ k \end{bmatrix}_q := \frac{(q)_m}{(q)_{m-k}(q)_k}, \quad (q)_k := \prod_{j=1}^k (1 - q^j) \tag{4.116}$$

with the convention $(q)_0 = 1$.

Exercise 4.9 Prove the q-binomial formula (4.115).

The coproduct of the basis elements of the algebra B_q is calculated as follows:

$$\Delta(b^m a^n) = \Delta(b)^m \Delta(a)^n = (a \otimes b + b \otimes 1)^m (a \otimes a)^n$$

$$= \sum_{s=0}^{m} \begin{bmatrix} m \\ s \end{bmatrix}_q (b \otimes 1)^{m-s} (a \otimes b)^s (a \otimes a)^n = \sum_{s=0}^{m} \begin{bmatrix} m \\ s \end{bmatrix}_q b^{m-s} a^{s+n} \otimes b^s a^n$$

$$(4.117)$$

where, in the third equality, we used the q-binomial formula (4.115) with $u = a \otimes b$ and $v = b \otimes 1$.

Exercise 4.10 Show that the antipode of B_q acts on the basis elements as follows

$$S(b^m a^n) = \left(-q^{-n-\frac{m+1}{2}}\right)^m b^m a^{-m-n}. \tag{4.118}$$

Let us now turn to the description of the restricted dual of B_q in the case when q is not a root of unity, that is $1 \notin q^{\mathbb{Z} \neq 0}$.

We start by remarking that there exist two morphisms of Hopf algebras

$$r : B_q \to \mathbb{C}[\mathbb{Z}] \simeq \mathbb{C}[t, t^{-1}], \quad a \mapsto t, \ b \mapsto 0, \tag{4.119}$$

and

$$s : \mathbb{C}[t, t^{-1}] \to B_q, \quad t \mapsto a, \tag{4.120}$$

which satisfy the relation $rs = \mathrm{id}_{\mathbb{C}[t,t^{-1}]}$ corresponding to the commutative diagram

$$\begin{array}{ccc}
\mathbb{C}[t, t^{-1}] & =\!\!=\!\!=\!\!=\!\!= & \mathbb{C}[t, t^{-1}] \\
& {}_s \searrow \quad {}^r \nearrow & \\
& B_q &
\end{array} \tag{4.121}$$

Thus, we have the induced commutative diagram for the restricted duals

$$\begin{array}{ccc}
\mathbb{C}[C_{\neq 0}] \otimes \mathbb{C}[x] & =\!\!=\!\!=\!\!=\!\!= & \mathbb{C}[C_{\neq 0}] \otimes \mathbb{C}[x] \\
& {}_{s^o} \nwarrow \quad {}^{r^o} \swarrow & \\
& B_q^o &
\end{array} \tag{4.122}$$

where we use the isomorphism of Hopf algebras

$$\mathbb{C}[t, t^{-1}]^o \simeq \mathbb{C}[C_{\neq 0}] \otimes \mathbb{C}[x] \tag{4.123}$$

with the evaluations

$$\langle \chi_z \otimes 1, t^n \rangle = z^n, \quad \langle 1 \otimes x, t^n \rangle = n, \quad \forall n \in \mathbb{Z}. \tag{4.124}$$

In particular, we see that r^o is injective and s^o is surjective.

The Hopf subalgebra $r^o(\mathbb{C}[\mathbb{C}_{\neq 0}] \otimes \mathbb{C}[x]) \subset B_q^o$ is generated by the grouplike elements $\theta_z := r^o(\chi_z \otimes 1)$, $z \in \mathbb{C}_{\neq 0}$, and the primitive element $\psi := r^o(1 \otimes x)$ determined by the formulae

$$\langle \theta_z, b^m a^n \rangle = \langle \chi_z \otimes 1, r(b^m a^n) = \delta_{m,0} \langle \chi_z \otimes 1, t^n \rangle = \delta_{m,0} z^n, \tag{4.125}$$

and

$$\langle \psi, b^m a^n \rangle = \langle 1 \otimes x, r(b^m a^n) \rangle = \delta_{m,0} \langle 1 \otimes x, t^n \rangle = \delta_{m,0} n \tag{4.126}$$

where $(m, n) \in \omega \times \mathbb{Z}$. Any $f \in B_q^o$ determines a (unique) set of polynomials

$$\{p_{f,z}(x)\}_{z \in J_f} \subset \mathbb{C}[x] \tag{4.127}$$

indexed by a finite set $J_f \subset \mathbb{C}_{\neq 0}$ such that

$$s^o f = \sum_{z \in J_f} \chi_z \otimes p_{f,z}(x) \implies (sr)^o f = \sum_{z \in J_f} \theta_z p_{f,z}(\psi). \tag{4.128}$$

Next, we consider the following two-dimensional representation:

$$\lambda : B_q \to \mathrm{End}(\mathbb{C}^2), \quad a \mapsto \begin{pmatrix} q & 0 \\ 0 & 1 \end{pmatrix}, \quad b \mapsto \begin{pmatrix} 0 & 1-q \\ 0 & 0 \end{pmatrix}. \tag{4.129}$$

Exercise 4.11 Show that

$$\lambda(b^m a^n) = \begin{pmatrix} \delta_{m,0} q^n & \delta_{m,1}(1-q) \\ 0 & \delta_{m,0} \end{pmatrix}, \quad \forall m \in \omega, \ n \in \mathbb{Z}, \tag{4.130}$$

The upper off-diagonal matrix coefficient of this representation $\phi := \lambda_{1,2}$ gives us a new element of B_q^o:

$$\langle \phi, b^m a^n \rangle = \delta_{m,1}(1-q), \quad \forall m \in \omega, \ n \in \mathbb{Z}, \tag{4.131}$$

with the coproduct

$$\Delta \phi = \Delta \lambda_{1,2} = \lambda_{1,1} \otimes \lambda_{1,2} + \lambda_{1,2} \otimes \lambda_{2,2} = \theta_q \otimes \phi + \phi \otimes 1 \tag{4.132}$$

where we use the equalities $\lambda_{1,1} = \theta_q$ and $\lambda_{2,2} = \theta_1 = 1$.

Exercise 4.12 Show that

$$\langle \phi^k, b^m a^n \rangle = \delta_{k,m}(q)_m, \quad \forall k, m \in \omega, \ n \in \mathbb{Z}, \tag{4.133}$$

see (4.116) for the definition of the symbol $(q)_m$.

Proposition 4.3 *The following relation is satisfied in B_q^o*

$$\psi\phi - \phi\psi = \phi. \tag{4.134}$$

Proof Let us calculate

$$\langle \psi\phi, b^m a^n \rangle = \langle \psi \otimes \phi, (a \otimes b + b \otimes 1)^m (a^n \otimes a^n) \rangle = \langle \psi \otimes \phi, a^{m+n} \otimes b^m a^n \rangle$$

$$= \langle \psi, a^{m+n} \rangle \langle \phi, b^m a^n \rangle = (m+n)\delta_{m,1}(q)_1 = \delta_{m,1}(n+1)(q)_1, \tag{4.135}$$

and

$$\langle \phi\psi, b^m a^n \rangle = \langle \phi \otimes \psi, (a \otimes b + b \otimes 1)^m (a^n \otimes a^n) \rangle = \langle \phi \otimes \psi, b^m a^n \otimes a^n \rangle$$

$$= \langle \phi, b^m a^n \rangle \langle \psi, a^n \rangle = n\delta_{m,1}(q)_1. \tag{4.136}$$

Thus,

$$\langle \psi\phi - \phi\psi, b^m a^n \rangle = \delta_{m,1}(q)_1 = \langle \phi, b^m a^n \rangle. \tag{4.137}$$

Exercise 4.13 Show that

$$\theta_z \phi = z\phi\theta_z, \quad \forall z \in \mathbb{C}_{\neq 0}. \tag{4.138}$$

Lemma 4.6 *Let $\lambda : B_q \to \mathrm{End}(V)$ be a finite dimensional representation of B_q. Then the element λb is nilpotent.*

Proof It suffices to prove that λb does not have non-zero eigenvalues. Indeed, assume the contrary, i.e. that there exists $\alpha \in \mathbb{C}_{\neq 0}$ and $v \in V_{\neq 0}$ such that

$$(\lambda b)v = \alpha v. \tag{4.139}$$

Then, for any $n \in \mathbb{Z}$, the vector $(\lambda a)^{-n} v$ is a non-trivial eigenvector corresponding to the eigenvalue $q^n \alpha$. Taking into account the fact that q is not a root of unity, the eigenvalues $\{q^n \alpha\}_{n \in \mathbb{Z}}$ are pairwise distinct. Thus, we come to the conclusion that the spectrum of λb is an infinite set which is impossible for a finite dimensional matrix. $\qquad \square$

Corollary 4.2 *For any $f \in B_q^o$, there exists $n \in \omega$ such that $\langle f, b^m \rangle = 0$ for any $m \geq n$.*

The following theorem summarises the Hopf algebra structure of the restricted dual of B_q.

Theorem 4.4 *Let $q \in \mathbb{C}$ be such that $1 \notin q^{\mathbb{Z} \neq 0}$. The Hopf algebra B_q^o is generated by the elements $\{\theta_z\}_{z \in \mathbb{C} \neq 0}$, ψ and ϕ defined in (4.125), (4.126) and (4.131) with the coproducts*

$$\Delta \theta_z = \theta_z \otimes \theta_z, \quad \Delta \psi = 1 \otimes \psi + \psi \otimes 1, \quad \Delta \phi = \theta_q \otimes \phi + \phi \otimes 1 \qquad (4.140)$$

and satisfying the relations

$$\theta_z \theta_w = \theta_{zw}, \quad \theta_z \psi = \psi \theta_z, \qquad (4.141)$$

$$\theta_z \phi = z \phi \theta_z, \quad \psi \phi - \phi \psi = \phi. \qquad (4.142)$$

The set $W := \{\phi^m \theta_z \psi^n \mid m, n \in \omega, \ z \in \mathbb{C}_{\neq 0}\}$ is a linear basis of B_q^o.

Proof The coproducts and the relations between the generating elements are already derived, see Theorem 4.2, Eqs. (4.129)–(4.132), Proposition 4.3, and Exercise 4.13. Let us show that any $f \in B_q^o$ is a finite linear combination of elements of the set W.

Let a finite subset $u \subset (B_q^o)^2$ be such that $\Delta f = \sum_{g \in u} g_0 \otimes g_1$. By Corollary 4.2, there exists $v \in \omega$ such that $\langle g_0, b^m \rangle = 0$ for any $g \in u$ and any $m \geq v$.

Denoting $\underline{v} := \{0, 1, \dots, v - 1\}$, define the element

$$\tilde{f} := \sum_{\substack{g \in u \\ k \in \underline{v}}} \frac{\langle g_0, b^k \rangle}{(q)_k} \phi^k (sr)^o g_1 \in B_q^o. \qquad (4.143)$$

The following calculation shows that $\tilde{f} = f$:

$$\langle \tilde{f}, b^m a^n \rangle = \sum_{\substack{g \in u \\ k \in \underline{v}}} \frac{\langle g_0, b^k \rangle}{(q)_k} \langle \phi^k (sr)^o g_1, b^m a^n \rangle$$

$$= \sum_{\substack{g \in u \\ k \in \underline{v}}} \frac{\langle g_0, b^k \rangle}{(q)_k} \langle \phi^k \otimes (sr)^o g_1, \Delta(b^m a^n) \rangle = \sum_{\substack{g \in u \\ k \in \underline{v}}} \frac{\langle g_0, b^k \rangle}{(q)_k} \langle \phi^k, b^m a^n \rangle \langle (sr)^o g_1, a^n \rangle$$

$$= \sum_{\substack{g \in u \\ k \in \underline{v}}} \frac{\langle g_0, b^k \rangle}{(q)_k} \delta_{k,m} (q)_k \langle g_1, a^n \rangle = \sum_{g \in u} \langle g_0, b^m \rangle \langle g_1, a^n \rangle$$

$$= \langle \Delta f, b^m \otimes a^n \rangle = \langle f, b^m a^n \rangle. \qquad (4.144)$$

In order to prove that the set W is linearly independent, assume that there exists a subset

$$\{f_i\}_{i \in I} \subset \mathbb{C}[t, t^{-1}]^o \simeq \mathbb{C}[\mathbb{C}_{\neq 0}] \otimes \mathbb{C}[x] \tag{4.145}$$

indexed by a finite subset $I \subset \omega$ such that $\sum_{i \in I} \phi^i r^o f_i = 0$. Evaluating at the basis elements $b^m a^n$ with $m \in I$, we obtain

$$
\begin{aligned}
0 = \sum_{i \in I} \langle \phi^i r^o f_i, b^m a^n \rangle &= \sum_{i \in I} \langle \phi^i \otimes r^o f_i, \Delta(b^m a^n) \rangle \\
&= \sum_{i \in I} \langle \phi^i \otimes r^o f_i, b^m a^n \otimes a^n \rangle = (q)_m \sum_{i \in I} \delta_{i,m} \langle r^o f_i, a^n \rangle = (q)_m \langle r^o f_m, a^n \rangle
\end{aligned}
\tag{4.146}
$$

which means that

$$\langle r^o f_m, a^n \rangle = \langle f_m, t^n \rangle = 0, \quad \forall (m, n) \in I \times \mathbb{Z}. \tag{4.147}$$

We conclude that $f_m = 0$ for any $m \in I$. $\qquad\qquad\square$

Chapter 5
The Quantum Double

The quantum double construction originally has been introduced by V. Drinfel'd in [11]. It allows to associate to any Hopf algebra with invertible antipode another Hopf algebra whose category of finite-dimensional representations is canonically braided. In this chapter, following [21], we describe the construction of the quantum double by using the notion of a cocycle over a bialgebra.

5.1 Bialgebras Twisted by Cocycles

Definition 5.1 A *cocycle* in a bialgebra $B = (B, \mu, \eta, \Delta, \epsilon)$ is an invertible element v of the convolution algebra $(B \otimes B)^*$ such that

$$v((v * \mu) \otimes \mathrm{id}_B) = v(\mathrm{id}_B \otimes (v * \mu)) \Leftrightarrow \qquad \text{(5.1)}$$

and

$$v(\eta \otimes \mathrm{id}_B) = \epsilon = v(\mathrm{id}_B \otimes \eta) \Leftrightarrow \qquad \text{(5.2)}$$

where $v * \mu := (\eta v) * \mu$ is the convolution product in the space of linear maps $L(B \otimes B, B)$.

Remark 5.1 Equation (5.1) can equivalently be written as the following identity in the convolution algebra $(B^{\otimes 3})^*$

$$v_{1,2} * (v(\mu \otimes \mathrm{id}_B)) = v_{2,3} * (v(\mathrm{id}_B \otimes \mu)). \qquad \text{(5.3)}$$

R. Kashaev, *A Course on Hopf Algebras*, Universitext,
https://doi.org/10.1007/978-3-031-26306-4_5

Exercise 5.1 Show that the convolution inverse $\bar{v}$ of a cocycle v in a bialgebra B satisfies the conditions

$$\bar{v}((\mu * \bar{v}) \otimes id_B) = \bar{v}(id_B \otimes (\mu * \bar{v})) \Leftrightarrow \qquad \text{(5.4)}$$

$$\bar{v}(\eta \otimes id_B) = \epsilon = \bar{v}(id_B \otimes \eta) \Leftrightarrow \qquad \text{(5.5)}$$

Exercise 5.2 Let H be a Hopf algebra. Define a linear form

$$v_H := \epsilon_H \otimes ev_H \otimes \epsilon_{H^o} = \qquad \in (H \otimes H^o \otimes H \otimes H^o)^* \qquad \text{(5.6)}$$

where

$$ev_H = \qquad : H^o \otimes H \to \mathbb{F}, \quad f \otimes x \mapsto \langle f, x \rangle, \qquad \text{(5.7)}$$

is the evaluation form. Show that this linear form is a cocycle in the bialgebra $H \otimes H^{o,\mathrm{op}}$, and the linear form

$$\bar{v}_H = \epsilon_H \otimes (ev_H(id_{H^o} \otimes S_H)) \otimes \epsilon_{H^o} = \qquad \text{(5.8)}$$

is its convolution inverse.

Proposition-Definition 5.1 *Let* $B = (B, \mu, \Delta, \eta, \epsilon)$ *be a bialgebra and* v *a cocycle in* B. *Then, the multiple* $B_v := (B, \mu_v, \Delta, \eta, \epsilon)$ *with the twisted product*

$$\mu_v := v * \mu * \bar{v} \qquad \text{(5.9)}$$

is a bialgebra called the bialgebra twisted by cocycle v. $\qquad \square$

Proof We have to check all the properties containing the product, i.e. the associativity, the unitality, the compatibility, and the compatibility of the product and the counit. The graphical notation

$$
\nu := \quad, \quad \bar{\nu} := \quad, \quad \mu_\nu := \quad = \quad \tag{5.10}
$$

allows us to proceed purely graphically as follows.

(1) Associativity:

$$
= \quad = \quad \tag{5.11}
$$

$$
= \quad =
$$

and

$$
= \quad = \quad \tag{5.12}
$$

We observe that the associativity for the twisted product is satisfied as a consequence of the cocycle relations (5.1) and (5.4). Notice that the diagrammatic calculations in (5.11) and (5.12) are mirror images of one another (with respect to a vertical mirror) accompanied with exchange of ν and $\bar{\nu}$.

(2) Unitality:

$$(5.13)$$

(3) Compatibility:

$$(5.14)$$

(4) Compatibility of the twisted product with the counit:

$$
\begin{array}{c} \end{array} \quad = \quad \begin{array}{c} \end{array} \quad = \quad \begin{array}{c} \end{array} \quad = \quad \begin{array}{c} \end{array}
$$

$$(5.15)$$

$$\square$$

5.1.1   Dual Pairings

The algebraic properties of the evaluation form given by relations (4.2) and (4.3) can be formalized into the notion of a dual pairing. One can construct cocycles as dual pairings possessing an extra property.

Definition 5.2 A *dual pairing* between two bialgebras A and B is a linear form $\varphi \in (A \otimes B)^*$ such that

$$\varphi(\mu_A \otimes \mathrm{id}_B) = \varphi_{13} * \varphi_{23} \Leftrightarrow$$

$$(5.16)$$

in the convolution algebra $(A \otimes A \otimes B)^*$ and

$$\varphi(\mathrm{id}_A \otimes \mu_B) = \varphi_{12} * \varphi_{13} \Leftrightarrow$$

$$(5.17)$$

in the convolution algebra $(A \otimes B \otimes B)^*$,

$$\varphi(\eta_A \otimes \mathrm{id}_B) = \epsilon_B \Leftrightarrow$$

$$(5.18)$$

and

$$\varphi(\mathrm{id}_A \otimes \eta_B) = \epsilon_A \Leftrightarrow$$

$$(5.19)$$

where, in the graphical notation, the dotted lines correspond to A and solid lines to B.

Proposition 5.1 *For any bialgebras A and B, a linear form $\varphi \in (A \otimes B)^*$ is a dual pairing between A and B if and only if one of the two following linear maps*

$$l: A \to B^*, \; r: B \to A^*, \quad \langle l(a), b \rangle = \langle r(b), a \rangle = \langle \varphi, a \otimes b \rangle, \qquad (5.20)$$

factorizes through a bialgebra homomorphism into the corresponding restricted dual.

Proof Assuming that φ is a dual pairing, we verify that $l(A) \subset B^o$ and $l: A \to B^o$ is a homomorphism of bialgebras. To this end, we first derive the equalities

$$\mu_B^* l = (l \otimes l) \Delta_A, \quad \eta_B^* l = \epsilon_A \qquad (5.21)$$

which imply that $l(A) \subset B^o$ and that l is a homomorphism of coalgebras. Using Sweedler's sigma notation for the coproduct, see Sect. 1.7.2,

$$\Delta(a) := \sum_{(a)} a_{(1)} \otimes a_{(2)}, \qquad (5.22)$$

for any $a \in A$ and $\alpha \otimes \beta \in B^{\otimes 2}$, we have

$$\langle \mu_B^*(l(a)), \alpha \otimes \beta \rangle = \langle l(a), \alpha\beta \rangle = \langle \varphi, a \otimes \alpha\beta \rangle$$
$$= \sum_{(a)} \langle \varphi, a_{(1)} \otimes \alpha \rangle \langle \varphi, a_{(2)} \otimes \beta \rangle = \sum_{(a)} \langle l(a_{(1)}), \alpha \rangle \langle l(a_{(2)}), \beta \rangle = \langle (l \otimes l)(\Delta_A(a)), \alpha \otimes \beta \rangle,$$

$$(5.23)$$

obtaining the first equality of (5.21), and

$$\eta_B^*(l(a)) = \langle l(a), \eta_B(1) \rangle = \langle \varphi, a \otimes \eta_B(1) \rangle = \epsilon_A(a), \qquad (5.24)$$

obtaining the second equality of (5.21).

Next, we show that

$$\Delta_B^*(l \otimes l) = l\mu_A, \quad l\eta_A = \epsilon_B^* \qquad (5.25)$$

which imply that l is a homomorphism of algebras. For any $a \otimes b \in A^{\otimes 2}$ and $\alpha \in B$, we have

$$\langle (l\mu_A)(a \otimes b), \alpha \rangle = \langle l(ab), \alpha \rangle = \langle \varphi, ab \otimes \alpha \rangle$$
$$= \sum_{(\alpha)} \langle \varphi, a \otimes \alpha_{(1)} \rangle \langle \varphi, b \otimes \alpha_{(2)} \rangle = \sum_{(\alpha)} \langle l(a), \alpha_{(1)} \rangle \langle l(b), \alpha_{(2)} \rangle = \langle l(a) \otimes l(b), \Delta_B(\alpha) \rangle$$

$$= \langle \Delta_B^*(l(a) \otimes l(b)), \alpha \rangle, \qquad (5.26)$$

obtaining the first equality of (5.25), and

$$\langle l(\eta_A(1)), \alpha \rangle = \langle \varphi, \eta_A(1) \otimes \alpha \rangle = \langle \epsilon_B, \alpha \rangle, \tag{5.27}$$

obtaining the second equality of (5.25).

Assuming now the converse, i.e. that Eqs. (5.21) and (5.25) are satisfied, the calculations of (5.23), (5.24), (5.26), (5.27) reproduce the definition of a dual pairing.

The case where l is replaced with r is checked similarly.

□

Proposition 5.2 *For any bialgebra B, a convolution invertible dual pairing φ between B^{op} and B (or, equivalently, between B and B^{cop}) is a cocycle on B if and only if*

$$\varphi_{12} * \varphi_{23} = \varphi_{23} * \varphi_{12} \tag{5.28}$$

in the convolution algebra $\left(B^{\otimes 3} \right)^*$.

Proof Relations (5.16) and (5.17) take the form

$$\varphi(\mu_B \otimes id_B) = \varphi_{23} * \varphi_{13} \tag{5.29}$$

and

$$\varphi(id_B \otimes \mu_B) = \varphi_{12} * \varphi_{13} \tag{5.30}$$

so that (5.3) takes the form

$$\varphi_{12} * \varphi_{23} * \varphi_{13} = \varphi_{23} * \varphi_{12} * \varphi_{13} \Leftrightarrow \varphi_{12} * \varphi_{23} = \varphi_{23} * \varphi_{12}. \tag{5.31}$$

□

5.2 Cobraided Bialgebras

Definition 5.3 A *dual universal r-matrix* in a bialgebra $B = (B, \mu, \Delta, \eta, \epsilon)$ is a convolution invertible element $\rho \in (B \otimes B)^*$ such that

$$\rho * \mu = \mu^{op} * \rho \Leftrightarrow \quad \tag{5.32}$$

$$\rho_{1,3} * \rho_{1,2} = \rho(\mathrm{id}_B \otimes \mu) \quad \Leftrightarrow \qquad \qquad \tag{5.33}$$

$$\rho_{1,3} * \rho_{2,3} = \rho(\mu \otimes \mathrm{id}_B) \quad \Leftrightarrow \qquad \qquad \tag{5.34}$$

A bialgebra provided with a dual universal r-matrix is called *cobraided*.

Exercise 5.3 Show that a dual universal r-matrix in a bialgebra B satisfies the following Yang–Baxter relation in the convolution algebra $\left(B^{\otimes 3}\right)^*$:

$$\rho_{1,2} * \rho_{1,3} * \rho_{2,3} = \rho_{2,3} * \rho_{1,3} * \rho_{1,2}. \tag{5.35}$$

5.2.1   The Quantum Double

In this subsection, a Hopf algebra H will be drawn graphically by solid lines while its restricted dual H^o by dotted lines.

Proposition-Definition 5.2 *Let H be a Hopf algebra. The* quantum double $D(H)$ *of H is the bialgebra $H \otimes H^{o,\mathrm{op}}$ twisted by the cocycle*

$$\nu_H = \epsilon_H \otimes \mathrm{ev}_H \otimes \epsilon_{H^o} = \qquad \qquad . \tag{5.36}$$

It contains bialgebras H and $H^{o,\mathrm{op}}$ as sub-bialgebras through the following canonical bialgebra embeddings:

$$\iota = \qquad = \qquad : H \hookrightarrow D(H), \quad x \mapsto x \otimes 1_{H^o}, \tag{5.37}$$

and

$$J = \qquad = \qquad : H^{o,\mathrm{op}} \hookrightarrow D(H), \quad f \mapsto 1_H \otimes f. \tag{5.38}$$

If the antipode of H is invertible, then $D(H)$ is a Hopf algebra.   □

Proof The proof boils down to straightforward verifications which are left as exercise. In particular, for the cocycle property, see Exercise 5.2. □

Exercise 5.4 Show that

$$\mu_{D(H)} = \quad \boxed{\mu_{D(H)}} \quad = \quad \boxed{\psi} \quad = (\mu_H \otimes \mu_{H^o}^{op})(\mathrm{id}_H \otimes \psi \otimes \mathrm{id}_{H^o})$$

$$(5.39)$$

where

$$\psi = \quad \boxed{\psi} \quad = \quad \text{}$$

$$= (\mathrm{ev}_H \otimes \mathrm{id}_{H \otimes H^o} \otimes \mathrm{ev}_H)(\mathrm{id}_{H^o} \otimes \sigma_{(H^o)^{\otimes 2}, H^{\otimes 2}} \otimes S_H)(\Delta_{H^o}^{(3)} \otimes \Delta_H^{(3)}) : H^o \otimes H \to H \otimes H^o,$$

$$\psi(f \otimes x) = \sum_{(f),(x)} \langle f_{(1)}, x_{(1)} \rangle x_{(2)} \otimes f_{(2)} \langle f_{(3)}, S(x_{(3)}) \rangle.$$

$$(5.40)$$

Theorem 5.1 *Let H be a Hopf algebra with invertible antipode. Then the restricted dual $D(H)^o$ of the quantum double $D(H)$ is a cobraided Hopf algebra with the following dual universal r-matrix*

$$\rho = \quad \boxed{\rho} \quad = \mathrm{ev}_{D(H)}\left(\mathrm{id}_{D(H)^o} \otimes (J\iota^o)\right) = \quad \text{}$$

$$(5.41)$$

with the convolution inverse

$$\bar{\rho} = \quad \boxed{\bar{\rho}} \quad = \mathrm{ev}_{D(H)}\left(\mathrm{id}_{D(H)^o} \otimes (J S_{H^o}^{-1} \iota^o)\right) = \quad \text{}$$

$$(5.42)$$

where we use thick lines for the restricted dual $D(H)^o$ and the graphical notation for the inverse of the antipode of H^o

$$S_{H^o}^{-1} := \quad \text{}.$$

Proof Let us see first that $\bar{\rho}$ is a right inverse of ρ

That $\bar{\rho}$ is a left inverse of ρ is verified similarly.

In order to verify equality (5.32), we write it in an equivalent graphical form

$$(5.43)$$

where the equivalence is due to the fact that two linear forms on a vector space are equal if and only if they evaluate to one and the same value on any vector.

By using the definitions of ρ and the product of $D(H)^o$, we rewrite Eq. (5.43) in the form

$$(5.44)$$

Next, we can use the definition of the coproduct of $D(H)^o$ in the bottom left parts of the diagrammatic equality (5.44) to obtain

$$(5.45)$$

where the units η_H can be eliminated by using the unitality axiom for H in the left hand side, and the definition of ψ in the right hand side

$$(5.46)$$

The obtained equality is a consequence of the equality (if two vectors are equal then their images by one and the same linear form are also equal)

$$(5.47)$$

which, in its turn, is equivalent to the equality (two linear forms on a vector space are equal if and only if they evaluate to one and the same value on any vector)

$$(5.48)$$

Now, in (5.48) we can use the definition of the product of H^o to obtain

$$(5.49)$$

By using the definitions of ψ in the left hand side and ι^o in the right hand side of (5.49), we obtain the equivalent equality

$$(5.50)$$

where we can further use the definitions of the (twice iterated) coproduct of H^o in the left hand side and the coproduct of $D(H)^o$ in the right hand side to obtain

$$(5.51)$$

In (5.51), we can use the definition of ι^o in the left hand side and the unitality axiom for H^o in the right hand side to obtain

$$(5.52)$$

In (5.52), we can use the definitions of the coproduct of $D(H)^o$ in the left hand side and of ψ in the right hand side to obtain

$$(5.53)$$

In the left hand side of (5.53), the definition of ψ and the composition of it with the unit of H^o lead to a simplification, while in the right hand side, the co-associativity properties of H and H^o and the duality allow to remove the antipode by the invertibility axiom. In this way, we obtain

$$\text{(5.54)}$$

In the left hand side of (5.54), the associativity and the co-associativity of H allow to remove the last antipode through the invertibility axiom for H. In this way, we obtain a tautological equality

$$\text{(5.55)}$$

Thus, equality (5.32) is proved.

Next, we verify equality (5.33)

$$\text{(5.56)}$$

Finally, we verify equality (5.34)

$$\qquad\qquad\qquad\qquad\qquad\qquad\qquad\qquad\qquad\qquad\qquad\qquad\qquad\qquad\qquad\qquad\qquad \square$$

Remark 5.2 If H is a finite-dimensional Hopf algebra with a linear basis $\{e_i\}_{i \in I}$ and $\{e^i\}_{i \in I}$ is the dual linear basis of H^*, then, the dual universal r-matrix is conjugate to the universal r-matrix

$$R := \sum_{i \in I} \jmath e^i \otimes \imath e_i \in D(H) \otimes D(H) \qquad (5.57)$$

in the sense that, for any $x, y \in (D(H))^o = (D(H))^*$, we have

$$\langle x \otimes y, R \rangle = \sum_{i \in I} \langle x, \jmath e^i \rangle \langle y, \imath e_i \rangle = \sum_{i \in I} \langle x, \jmath e^i \rangle \langle \imath^o y, e_i \rangle$$

$$= \left\langle x, \jmath \left(\sum_{i \in I} \langle \imath^o y, e_i \rangle e^i \right) \right\rangle = \langle x, \jmath \imath^o y \rangle = \langle \varrho, x \otimes y \rangle. \qquad (5.58)$$

In the infinite-dimensional case, formula (5.57) is formal but it is a convenient and useful tool for actual calculations.

5.3   The Quantum Double $D(B_q)$

In this section, we consider the example of the quantum group B_q described in Sect. 4.4 of Chap. 4. Recall that the parameter q there is generic, that it is not a root of unity.

Proposition 5.3 *Let* $q \in \mathbb{C}_{\neq 0}$ *be such that* $1 \notin q^{\mathbb{Z}_{\neq 0}}$. *Then, the quantum double* $D(B_q)$ *admits the following presentation:*

$$\mathbb{C}\langle a, b, \psi, \phi, \{\theta_z\}_{z \in \mathbb{C}_{\neq 0}} \mid ab = qba,$$

$$\psi\theta_z = \theta_z\psi, \ \theta_z\theta_w = \theta_{zw}, \ \phi\psi - \psi\phi = \phi, \ \phi\theta_z = z\theta_z\phi,$$

$$\psi a = a\psi, \ \psi b - b\psi = b, \ \theta_z a = a\theta_z, \ \theta_z b = zb\theta_z,$$

$$\phi a = qa\phi, \ \phi b - qb\phi = (1-q)(1-a\theta_q);$$

$$\Delta a = a \otimes a, \ \Delta b = a \otimes b + b \otimes 1,$$

$$\Delta\psi = \psi \otimes 1 + 1 \otimes \psi, \ \Delta\theta_z = \theta_z \otimes \theta_z, \ \Delta\phi = \theta_q \otimes \phi + \phi \otimes 1\rangle \qquad (5.59)$$

Proof As the first two lines and the last two lines in the presentation are just the presentations of the Hopf sub-algebras B_q and $B_q^{o,\text{op}}$ put together, we need to check only the relations in the third and forth lines. These are relations between the generators of B_q and $B_q^{o,\text{op}}$ which are of the form

$$(\jmath f)(\imath x) = \sum_{(f),(x)} \langle f_{(1)}, x_{(1)}\rangle (\imath x_{(2)})(\jmath f_{(2)})\langle f_{(3)}, Sx_{(3)}\rangle, \quad x \in B_q, \ f \in B_q^{o,\text{op}}.$$

$$(5.60)$$

By writing informally just x instead of $\imath x$ and f instead of $\jmath f$, let us write out these relations one after another for $x \in \{a, b\}$ and $f \in \{\psi, \theta_z, \phi\}$ by using the iterated coproducts

$$\Delta^{(3)}a = a \otimes a \otimes a, \quad \Delta^{(3)}b = b \otimes 1 \otimes 1 + a \otimes b \otimes 1 + a \otimes a \otimes b \qquad (5.61)$$

and

$$\Delta^{(3)}\psi = \psi \otimes \epsilon \otimes \epsilon + \epsilon \otimes \psi \otimes \epsilon + \epsilon \otimes \epsilon \otimes \psi, \quad \Delta^{(3)}\theta_z = \theta_z \otimes \theta_z \otimes \theta_z,$$

$$\Delta^{(3)}\phi = \phi \otimes \epsilon \otimes \epsilon + \theta_q \otimes \phi \otimes \epsilon + \theta_q \otimes \theta_q \otimes \phi. \qquad (5.62)$$

The Case $(f = \psi, x = a)$ The first coproducts in (5.61) and (5.62) imply that relation (5.60) takes the form

$$\psi a = \langle \psi, a\rangle a\langle \epsilon, a^{-1}\rangle + \langle \epsilon, a\rangle a\psi\langle \epsilon, a^{-1}\rangle + \langle \epsilon, a\rangle a\langle \psi, a^{-1}\rangle = a + a\psi - a = a\psi.$$

$$(5.63)$$

The Case $(f = \psi, x = b)$ The second coproduct in (5.61) and the first one in (5.62) imply that

$$\psi b = \langle \psi, b\rangle 1\langle \epsilon, 1\rangle + \langle \psi, a\rangle b\langle \epsilon, 1\rangle + \langle \psi, a\rangle a\langle \epsilon, -a^{-1}b\rangle$$

$$+ \langle \epsilon, b\rangle \psi\langle \epsilon, 1\rangle + \langle \epsilon, a\rangle b\psi\langle \epsilon, 1\rangle + \langle \epsilon, a\rangle a\psi\langle \epsilon, -a^{-1}b\rangle$$

$$\langle \epsilon, b\rangle 1\langle \psi, 1\rangle + \langle \epsilon, a\rangle b\langle \psi, 1\rangle + \langle \epsilon, a\rangle a\langle \psi, -a^{-1}b\rangle$$

$$= (0 + b + 0) + (0 + b\psi + 0) + (0 + 0 + 0) = b + b\psi. \qquad (5.64)$$

The Case $(f = \theta_z, x = a)$ The first coproduct in (5.61) and the second one in (5.62) imply that

$$\theta_z a = \langle \theta_z, a\rangle a\theta_z\langle \theta_z, a^{-1}\rangle = za\theta_z z^{-1} = a\theta_z. \qquad (5.65)$$

The Case $(f = \theta_z, x = b)$ The second coproducts in (5.61) and (5.62) imply that

$$\theta_z b = \langle \theta_z, b \rangle \theta_z \langle \theta_z, 1 \rangle + \langle \theta_z, a \rangle b \theta_z \langle \theta_z, 1 \rangle + \langle \theta_z, a \rangle a \theta_z \langle \theta_z, -a^{-1}b \rangle$$

$$= 0 + zb\theta_z + 0 = zb\theta_z. \qquad (5.66)$$

The Case $(f = \phi, x = a)$ The first coproduct in (5.61) and the third one in (5.62) imply that

$$\phi a = \langle \phi, a \rangle a \langle \epsilon, a^{-1} \rangle + \langle \theta_q, a \rangle a \phi \langle \epsilon, a^{-1} \rangle + \langle \theta_q, a \rangle a \theta_q \langle \phi, a^{-1} \rangle$$

$$= 0 + qa\phi + 0 = qa\phi. \qquad (5.67)$$

The Case $(f = \phi, x = b)$ The second coproduct in (5.61) and the third one in (5.62) imply that

$$\phi b = \langle \phi, b \rangle 1 \langle \epsilon, 1 \rangle + \langle \phi, a \rangle b \langle \epsilon, 1 \rangle + \langle \phi, a \rangle a \langle \epsilon, -a^{-1}b \rangle$$

$$+ \langle \theta_q, b \rangle \phi \langle \epsilon, 1 \rangle + \langle \theta_q, a \rangle b \phi \langle \epsilon, 1 \rangle + \langle \theta_q, a \rangle a \phi \langle \epsilon, -a^{-1}b \rangle$$

$$+ \langle \theta_q, b \rangle \theta_q \langle \phi, 1 \rangle + \langle \theta_q, a \rangle b \theta_q \langle \phi, 1 \rangle + \langle \theta_q, a \rangle a \theta_q \langle \phi, -a^{-1}b \rangle$$

$$= ((1 - q)1 + 0 + 0) + (0 + qb\phi + 0) + (0 + 0 + qa\theta_q \langle \theta_q, -a^{-1} \rangle \langle \phi, b \rangle)$$

$$= (1 - q)1 + qb\phi - a\theta_q(1 - q) = (1 - q)(1 - a\theta_q) + qb\phi. \qquad (5.68)$$

□

5.3.1 Irreducible Representations of $D(B_q)$

Proposition 5.4 *The elements* $c, d \in D(B_q)$ *defined by the relations*

$$c\theta_q = a \qquad (5.69)$$

and

$$\phi b - 1 - qa\theta_q = \theta_q d = qb\phi - q - a\theta_q \qquad (5.70)$$

are central.

Proof That the element c is central is an easy check. To see that d is central, we define two elements $w, w' \in D(B_q)$ by the relations

$$\phi b = u + va\theta_q + w, \qquad qb\phi = u' + v'a\theta_q + w', \qquad (5.71)$$

where $u, u', v, v' \in \mathbb{C}$ are fixed as follows. First, we impose two conditions

$$u - u' = 1 - q = v' - v \tag{5.72}$$

which, due to the defining relation between b and ϕ, imply that $w' = w$. By straightforward verifications one sees that w commutes with a, ψ and θ_z for all $z \in \mathbb{C}_{\neq 0}$. Next, we have the equalities

$$u'b + v'a\theta_q b + wb = qb\phi b = qbu + qbva\theta_q + qbw \tag{5.73}$$

which, under two more relations of the form

$$u' = qu, \quad qv' = v, \tag{5.74}$$

imply that $wb = qbw$. The system of Eqs. (5.72) and (5.74) on unknowns u, u', v, v' admits a unique solution

$$u = 1 = v', \quad u' = q = v. \tag{5.75}$$

Now, it is an easy check that $\phi w = qw\phi$. Indeed, we have

$$\phi w = \phi(qb\phi - q - a\theta_q) = q\phi b\phi - q\phi - \phi a\theta_q$$
$$= q\phi b\phi - q\phi - q^2 a\theta_q\phi = q(\phi b - 1 - qa\theta_q)\phi = qw\phi. \tag{5.76}$$

Finally, the equality $w = \theta_q d$, together with the obtained commutation relations for w, implies that d is central. □

Proposition 5.5 *Let $q \in \mathbb{C}_{\neq 0}$ be such that $1 \notin q^{\mathbb{Z}_{\neq 0}}$. The center of the algebra $D(B_q)$ coincides with the polynomial subalgebra $\mathbb{C}[c, c^{-1}, d]$ where c and d are defined in (5.69) and (5.70)*

Proof By Proposition 5.4, for any $n \in \omega$, one can easily verify by recurrence the equality

$$\phi^n b^n = \prod_{k \in n} (1 + q^k \theta_q d + q^{2k+1} \theta_q^2 c). \tag{5.77}$$

This means that any element $x \in D(B_q)$ can uniquely be written in the form

$$x = \sum_{(u,m) \in \mathbb{C}_{\neq 0} \times \mathbb{Z}} \theta_u e_m p_{u,m}(c, d, \psi), \tag{5.78}$$

where

$$e_m := \begin{cases} b^m & \text{if } m > 0; \\ 1 & \text{if } m = 0; \\ \phi^{-m} & \text{if } m < 0 \end{cases} \tag{5.79}$$

and $p_{u,m}(a, c, \psi) \in \mathbb{C}[c, c^{-1}, d, \psi]$ is non-zero for only finitely many pairs (u, m). Remark that, for any $m \in \mathbb{Z}$, the element e_m satisfies the relations

$$\psi e_m = e_m(\psi + m), \quad \theta_z e_m = z^m e_m \theta_z \quad \forall z \in \mathbb{C}_{\neq 0}. \tag{5.80}$$

Assume that x is central. Then, for any $z \in \mathbb{C}_{\neq 0}$, we have the equality

$$x = \theta_z x \theta_z^{-1} = \sum_{(u,m) \in \mathbb{C}_{\neq 0} \times \mathbb{Z}} \theta_u e_m z^m p_{u,m}(c, d, \psi) \tag{5.81}$$

which implies that for any fixed pair $(u, m) \in \mathbb{C}_{\neq 0} \times \mathbb{Z}$, one has the family of equalities

$$p_{u,m} = z^m p_{u,m} \quad \forall z \in \mathbb{C}_{\neq 0}. \tag{5.82}$$

This means that $p_{u,m}$ can only be non-zero if $m = 0$. Thus, the element x takes the form

$$x = \sum_{u \in \mathbb{C}_{\neq 0}} \theta_u p_{u,0}(c, d, \psi). \tag{5.83}$$

The equality

$$bx = xb = b \sum_{u \in \mathbb{C}_{\neq 0}} \theta_u u p_{u,0}(a, c, \psi + 1) \tag{5.84}$$

is equivalent to the equalities

$$u p_{u,0}(c, d, \psi + 1) = p_{u,0}(c, d, \psi) \quad \forall u \in \mathbb{C}_{\neq 0} \tag{5.85}$$

which imply that the polynomial $p_{u,0}(a, c, \psi)$ can be non-zero only if $u = 1$ and if it does not depend on ψ. We conclude that $x = p_{1,0}(c, d) \in \mathbb{C}[c, c^{-1}, d]$. $\square$

Theorem 5.2 *Let $q \in \mathbb{C}$ be such that $1 \notin q^{\mathbb{Z}_{\neq 0}}$. Then, any finite dimensional irreducible representation $\lambda \colon D(B_q) \to \text{End}(V)$ is characterized by the dimension $N := \dim(V) \in \mathbb{Z}_{>0}$, a complex number $\gamma \in \mathbb{C}$, and a multiplicative group* *homomorphism* $\xi \colon \mathbb{C}_{\neq 0} \to \mathbb{C}_{\neq 0}$ *such that there exists a linear basis* $\{v_n\}_{n \in \underline{N}}$ *of* V *satisfying the relations*

$$(\lambda a) v_n = q^{N-1-n} \xi_q^{-1} v_n, \quad (\lambda \psi) v_n = (\gamma - n) v_n, \quad (\lambda \theta_z) v_n = z^{-n} \xi_z v_n,$$

$$(\lambda b) v_n = (1 - q^{-n}) v_{n-1}, \quad (\lambda \phi) v_n = (1 - q^{N-n-1}) v_{n+1}. \qquad (5.86)$$

Proof To simplify notation, we will write $\hat{x}$ instead of λx for any $x \in D(B_q)$, and $[x, y]$ instead of $xy - yx$.

As in an irreducible representation all central elements are realised by scalars, there exist $\alpha \in \mathbb{C}_{\neq 0}$ and $\beta \in \mathbb{C}$ such that the central elements c and d defined in (5.69) and (5.70) are represented by scalar multiples of the identity operator:

$$\hat{c} = \alpha \, \mathrm{id}_V, \quad \hat{d} = \beta \, \mathrm{id}_V. \qquad (5.87)$$

Let $u' \in V \setminus \{0\}$ be an eigenvector of $\hat{\psi}$ corresponding to an eigenvalue $\gamma' \in \mathbb{C}$. Then, the vector $\hat{b} u'$ either vanishes or it is an eigenvector of $\hat{\psi}$ corresponding to the eigenvalue $\gamma' + 1$. Indeed,

$$\hat{\psi} \hat{b} u' = ([\hat{\psi}, \hat{b}] + \hat{b} \hat{\psi}) u' = \hat{b}(1 + \hat{\psi}) u' = (\gamma' + 1) \hat{b} u'. \qquad (5.88)$$

Iterating the action of $\hat{b}$ and taking into account the fact that $\dim(V) < \infty$, we conclude that there exists a positive integer K such that $u'' := \hat{b}^{K-1} u' \neq 0$ and

$$\hat{b} u'' = 0, \quad \hat{\psi} u'' = \gamma u'', \quad \gamma := \gamma' + K - 1. \qquad (5.89)$$

Additionally, as the elements $\{\theta_z\}_{z \in \mathbb{C}_{\neq 0}}$ and ψ generate a commutative sub-algebra A of $D(B_q)$, and any irreducible finite dimensional representation of a commutative algebra is one dimensional, there exists a non zero vector $u \in \lambda(A) u''$ that generates an irreducible sub-representation of A. This means that the following relations are satisfied:

$$\hat{b} u = 0, \quad \hat{\psi} u = \gamma u, \quad \hat{\theta}_z u = \xi_z u, \quad \forall z \in \mathbb{C}_{\neq 0}, \qquad (5.90)$$

where

$$\xi \colon \mathbb{C}_{\neq 0} \to \mathbb{C}_{\neq 0} \qquad (5.91)$$

is a (multiplicative) group homomorphism.

By a similar reasoning, as in the case of the vector u' above, for any $n \in \omega$, the vector $\hat{\phi}^n u$ either vanishes or it is an eigenvector of $\hat{\psi}$ corresponding to the eigenvalue $\gamma - n$, and, as $\dim(V) < \infty$, there exists a positive integer M such that

$$\hat{\phi}^{M-1} u \neq 0, \quad \hat{\phi}^M u = 0. \qquad (5.92)$$

We denote by W the linear span of the vectors $\{\hat{\phi}^n u\}_{n \in \underline{M}}$. Let us show that $W = V$.

First, we note that, apart from the relations

$$\hat{\psi}\hat{\phi}^n u = (\gamma - n)\hat{\phi}^n u, \quad n \in \underline{M}, \tag{5.93}$$

we also have

$$\check{\theta}_z \hat{\phi}^n u = z^{-n} \hat{\phi}^n u, \quad n \in \underline{M}, \tag{5.94}$$

where we have denoted

$$\check{\theta}_z := \hat{\theta}_z / \xi_z, \quad \forall z \in \mathbb{C}_{\neq 0}. \tag{5.95}$$

Next, by using (5.87) in (5.70), we obtain

$$\hat{\phi}\hat{b} = \mathrm{id}_V + \beta\hat{\theta}_q + q\alpha\hat{\theta}_q^2 \tag{5.96}$$

and

$$\hat{b}\hat{\phi} = \mathrm{id}_V + q^{-1}\beta\hat{\theta}_q + q^{-1}\alpha\hat{\theta}_q^2 \tag{5.97}$$

Applying (5.96) to u and (5.97) to $\hat{\phi}^{M-1}u$, and taking into account relations (5.90), (5.92) and (5.94), we obtain

$$(1 + \beta\xi_q + q\alpha\xi_q^2)u = 0 \Rightarrow 1 + \beta\xi_q + \alpha q\xi_q^2 = 0 \tag{5.98}$$

and

$$(1 + \beta q^{-M}\xi_q + \alpha q^{1-2M}\xi_q^2)\hat{\phi}^{M-1}u = 0 \Rightarrow 1 + \beta q^{-M}\xi_q + \alpha q^{1-2M}\xi_q^2 = 0. \tag{5.99}$$

Excluding β from (5.98) and (5.99), we obtain

$$(1 - \alpha q^{1-M}\xi_q^2)(1 - q^M) = 0 \Leftrightarrow \alpha = q^{M-1}\xi_q^{-2} \tag{5.100}$$

and also from (5.98) it follows that

$$\beta = -\xi_q^{-1}(1 + q^M). \tag{5.101}$$

By using substitutions (5.100), (5.101) and notation (5.95), we rewrite (5.96) and (5.97) as follows:

$$\hat{\phi}\hat{b} = \mathrm{id}_V - (1 + q^M)\check{\theta}_q + q^M\check{\theta}_q^2 = (\mathrm{id}_V - \check{\theta}_q)(\mathrm{id}_V - q^M\check{\theta}_q) \tag{5.102}$$

and

$$\hat{b}\hat{\phi} = \mathrm{id}_V - (1 + q^M)q^{-1}\check{\theta}_q + q^{M-2}\check{\theta}_q^2 = (\mathrm{id}_V - q^{-1}\check{\theta}_q)(\mathrm{id}_V - q^{M-1}\check{\theta}_q). \tag{5.103}$$

For $n \in \underline{M} \setminus \{0\}$, applying relation (5.103) to the vector $\hat{\phi}^{n-1}u$ and taking into account (5.94), we obtain

$$\hat{b}\hat{\phi}^n u = (1 - q^{-n})(1 - q^{M-n})\hat{\phi}^{n-1}u. \tag{5.104}$$

Thus, we conclude that the subspace W of V generated by vectors $\{\hat{\phi}^n u\}_{n \in \underline{M}}$ is an invariant subspace of the representation λ, and by the irreducibility of λ, we conclude that $W = V$ so that

$$N := \dim(V) = \dim(W) = M, \tag{5.105}$$

and the vectors $\{\hat{\phi}^n u\}_{n \in \underline{M}}$ form a linear basis of V.

Let us define renormalized vectors

$$v_n := (q)_{N-n-1}\hat{\phi}^n u, \quad n \in \underline{N}. \tag{5.106}$$

Then, by using the relation

$$(1 - q^k)(q)_{k-1} = (q)_k, \quad \forall k \in \mathbb{Z}_{>0}, \tag{5.107}$$

we have

$$\hat{b}v_n = (q)_{N-n-1}(1 - q^{-n})(1 - q^{N-n})\hat{\phi}^{n-1}u = (1 - q^{-n})v_{n-1} \tag{5.108}$$

and

$$\hat{\phi}v_n = (q)_{N-n-1}\hat{\phi}^{n+1}v = (1 - q^{N-n-1})v_{n+1}. \tag{5.109}$$

$\square$

Remark 5.3 The vanishing properties of the coefficients of relations (5.108) with $n = 0$ and (5.109) with $n = N - 1$ naturally take care of the annihilation relations

$$\hat{b}v_0 = \hat{\phi}v_{N-1} = 0. \tag{5.110}$$

Exercise 5.5 For any $n \in \underline{N}$, show that

$$\hat{b}^k v_n = (q^{-n}; q)_k v_{n-k}, \quad \forall k \in \underline{n+1}, \tag{5.111}$$

and

$$\hat{\phi}^k v_n = (q^{N-n-k}; q)_k v_{n+k}, \quad \forall k \in \underline{N - n}. \tag{5.112}$$

with the notation

$$(x; q)_n := \begin{cases} \prod_{k=0}^{n-1}(1 - xq^k) & \text{if } k > 0; \\ 1 & \text{if } k = 0. \end{cases} \tag{5.113}$$

5.3.2   Quantum Group $U_q(sl_2)$

Recall that the element $c := a\theta_q^{-1} \in D(B_q)$ is central and grouplike. This means that the vector subspace

$$I_q := (c - 1)D(B_q) \subset D(B_q)$$

is a bi-ideal stable under the action of the antipode, see Definitions 2.6, 2.2 and 2.4. By the results of Chap. 2, Sect. 2.4.2, we conclude that the quotient vector space

$$H_q := D(B_q)/I_q$$

admits a unique structure of a Hopf algebra such that the canonical projection map $\pi: D(B_q) \to H_q$ is a morphism of Hopf algebras. The Hopf algebra H_q is closely related with the *quantum group* $U_q(sl_2)$ which is defined by the following presentation:

generators: k, e, f;

relations: $ke = q^2 ek, \quad kf = q^{-2} fk, \quad ef - fe = \dfrac{k - k^{-1}}{q - q^{-1}}$

coproducts: $\Delta k = k \otimes k, \quad \Delta e = k \otimes e + e \otimes 1, \quad \Delta f = 1 \otimes f + f \otimes k^{-1}$

where we assume that $q^2 \neq 1$ and k is invertible (as a group-like element in any Hopf algebra).

Exercise 5.6 Determine $\alpha, \beta \in \mathbb{C}_{\neq 0}$ such that the map

$$k \mapsto a + I_{q^2}, \quad e \mapsto \alpha b + I_{q^2}, \quad f \mapsto \beta a^{-1}\phi + I_{q^2}$$

extends to an injective morphism of Hopf algebras $h: U_q(sl_2) \to H_{q^2}$.

The algebra $U_q(sl_2)$ was discovered in [24], and the general theory of quantum groups has been subsequently developed in the works [11, 13, 17]. An introduction for this subject can be found in the book [16].

5.4 The Hopf Algebra $D(B_1)$

Let B_1 be the commutative Hopf algebra over $\mathbb{C}$ corresponding to the quantum group B_q with $q = 1$ defined and analyzed in Sect. 4.4 of Chap. 4 in the case of generic q, that is when q is not a root of unity. Here, we consider the case of the simplest root of unity $q = 1$. This Hopf algebra coincides with J_0, the specification of $J_\hbar$ to $\hbar = 0$, see Sect. 4.3 of Chap. 4. In Sect. 6.5 of Chap. 6, this algebra will be used for interpretation of the Alexander polynomial of knots as an example of a universal invariant. For this reason, below we briefly describe the restricted dual and the quantum double of B_1, leaving the detailed analysis to exercises.

5.4.1 The Restricted Dual Hopf Algebra $B_1^{o,\mathrm{op}}$

The opposite $B_1^{o,\mathrm{op}}$ of the restricted dual Hopf algebra B_1^o is composed of two Hopf subalgebras: the group algebra $\mathbb{C}[\mathrm{Aff}_1(\mathbb{C})]$ generated by group-like elements

$$\chi_{u,v}, \quad (u, v) \in \mathbb{C} \times \mathbb{C}_{\neq 0}, \quad \chi_{u,v}\chi_{u',v'} = \chi_{u+vu',vv'}, \tag{5.114}$$

and the universal enveloping algebra $U(\mathrm{Lie\,Aff}_1(\mathbb{C}))$ generated by two primitive elements ψ and ϕ satisfying the relation

$$\phi\psi - \psi\phi = \phi. \tag{5.115}$$

The relations between the generators of $\mathbb{C}[\mathrm{Aff}_1(\mathbb{C})]$ and $U(\mathrm{Lie\,Aff}_1(\mathbb{C}))$ are of the form

$$[\chi_{u,v}, \psi] = u\phi\chi_{u,v}, \quad \chi_{u,v}\phi = v\phi\chi_{u,v} \quad \forall(u, v) \in \mathbb{C} \times \mathbb{C}_{\neq 0} \tag{5.116}$$

where $[x, y] := xy - yx$. As linear forms on B_1, they are defined by the relations

$$\langle\chi_{u,v}, b^m a^n\rangle = u^m v^{-m-n},$$

$$\langle\phi, b^m a^n\rangle = \delta_{m,1}, \quad \langle\psi, b^m a^n\rangle = \delta_{m,0}n, \quad \forall(m, n) \in \mathbb{Z}_{\geq 0} \times \mathbb{Z}. \tag{5.117}$$

Exercise 5.7 By using the methods of Chap. 4, provide the details of the above description of the structure of the Hopf algebra $B_1^{o,\mathrm{op}}$.

5.4.2 The Quantum Double $D(B_1)$

The commutation relations (5.60) in the case of the quantum double $D(B_1)$ take the form

$$[\psi, b] = b, \quad [\phi, b] = 1 - a,$$

$$b\chi_{u,v} = \chi_{u,v}(bv + (a-1)u) \quad \forall(u,v) \in \mathbb{C} \times \mathbb{C}_{\neq 0} \tag{5.118}$$

and a is central.

Exercise 5.8 Prove the defining relations of $D(B_1)$ given by Eq. (5.118).

Exercise 5.9 Show that in any finite-dimensional representation of the algebra $D(B_1)$, the elements $1 - a$, b and ϕ are nilpotent.

The formal universal r-matrix of $D(B_1)$, see Remark 5.2, is given by the formula

$$R := (1 \otimes a)^{\psi \otimes 1} e^{\phi \otimes b} = \sum_{m,n \geq 0} \frac{1}{n!} \binom{\psi}{m} \phi^n \otimes (a-1)^m b^n \tag{5.119}$$

and it is well defined in the context of finite-dimensional representations for the following reason.

Any finite dimensional right comodule V over $(D(B_1))^o$ is a left module over $D(B_1)$ defined by

$$xv = \sum_{(v)} v_{(0)} \langle v_{(1)}, x \rangle, \quad \forall(x, v) \in D(B_1) \times V \tag{5.120}$$

where we extend Sweedler's sigma notation to comodules. Thus, it suffices to make sense of formula (5.119) in the case of an arbirary finite-dimensional representation of $D(B_1)$ where the elements $1 - a$, b and ϕ are necessarily nilpotent, so that the formal infinite double sum truncates to a well defined finite sum.

5.4.3 The Center of $D(B_1)$

Proposition 5.6 *The center of the algebra $D(B_1)$ is the polynomial subalgebra* $\mathbb{C}[a^{\pm 1}, c]$ *where*

$$c := \phi b + (a-1)\psi. \tag{5.121}$$

Proof It is easily verified that c is central. Any element $x \in D(B_1)$ can uniquely be written in the form

$$x = \sum_{(u,v,m) \in \mathbb{C} \times \mathbb{C}_{\neq 0} \times \mathbb{Z}} \chi_{u,v} e_m \, p_{u,v,m}(a, c, \psi), \tag{5.122}$$

where

$$e_m := \begin{cases} b^m & \text{if } m > 0; \\ 1 & \text{if } m = 0; \\ \phi^{-m} & \text{if } m < 0 \end{cases} \tag{5.123}$$

and the polynomial $p_{u,v,m}(a, c, \psi) \in \mathbb{C}[a^{\pm 1}, c, \psi]$ is non-zero for only finitely many triples (u, v, m).

Assume that $x \in D(B_1)$ is a central element. Then, for any $s \in \mathbb{C}_{\neq 0}$, we have the equality

$$x = \chi_{0,s}^{-1} x \chi_{0,s} = \sum_{(u,v,m) \in \mathbb{C} \times \mathbb{C}_{\neq 0} \times \mathbb{Z}} \chi_{u/s,v} e_m s^m \, p_{u,v,m}(a, c, \psi)$$

$$= \sum_{(u,v,m) \in \mathbb{C} \times \mathbb{C}_{\neq 0} \times \mathbb{Z}} \chi_{u,v} e_m s^m \, p_{us,v,m}(a, c, \psi) \tag{5.124}$$

which implies that for any fixed triple $(u, v, m) \in \mathbb{C} \times \mathbb{C}_{\neq 0} \times \mathbb{Z}$, one has the family of equalities

$$p_{u,v,m} = s^m \, p_{us,v,m} \quad \forall s \in \mathbb{C}_{\neq 0}. \tag{5.125}$$

This means that $p_{u,v,m}$ can only be non-zero if $u = m = 0$. Thus, the element x takes the form

$$x = \sum_{v \in \mathbb{C}_{\neq 0}} \chi_{0,v} p_{0,v,0}(a, c, \psi). \tag{5.126}$$

The equality

$$bx = xb = b \sum_{v \in \mathbb{C}_{\neq 0}} \chi_{0,v} v^{-1} p_{0,v,0}(a, c, \psi + 1). \tag{5.127}$$

is equivalent to the equalities

$$p_{0,v,0}(a, c, \psi + 1) = v^{-1} p_{0,v,0}(a, c, \psi) \quad \forall v \in \mathbb{C}_{\neq 0} \tag{5.128}$$

which imply that the polynomial $p_{0,v,0}(a, c, \psi)$ can be non-zero only if $v = 1$ and if it does not depend on ψ. We conclude that $x \in \mathbb{C}[a^{\pm 1}, c]$. $\square$

5.5 Solutions of the Yang–Baxter Equation

Definition 5.4 An *r-matrix over a coalgebra* C is an invertible element ρ of the convolution algebra $(C^{\otimes 2})^*$ such that the following Yang–Baxter equation is satisfied in the convolution algebra $(C^{\otimes 3})^*$:

$$\rho_{1,2} * \rho_{1,3} * \rho_{2,3} = \rho_{2,3} * \rho_{1,3} * \rho_{1,2}. \tag{5.129}$$

Example 5.1 The dual universal r-matrix of a cobraided bialgebra B is an r-matrix over the underlying coalgebra of B. $\square$

Definition 5.5 An *r-matrix over a vector space* V is an element $r \in \mathrm{Aut}(V^{\otimes 2})$ such that the following Yang–Baxter equation is satisfied in the algebra $\mathrm{End}(V^{\otimes 3})$:

$$r_{1,2} r_{2,3} r_{1,2} = r_{2,3} r_{1,2} r_{2,3}, \quad r_{1,2} := r \otimes \mathrm{id}_V, \quad r_{2,3} := \mathrm{id}_V \otimes r. \tag{5.130}$$

By using the graphical notation , the Yang–Baxter equation (5.130) takes the following graphical form

$$\tag{5.131}$$

In the particular case, where V is a finite dimensional vector space over a field $\mathbb{F}$, let $B \subset V$ be a linear basis. Defining the matrix coefficients

$$\{r_{a,b}^{c,d} \mid a, b, c, d \in B\} \subset \mathbb{F}, \quad r(a \otimes b) = \sum_{c,d \in B} r_{a,b}^{c,d} c \otimes d, \quad a, b \in B, \tag{5.132}$$

we reduce the Yang–Baxter equation (5.131) to a over determined system of non-linear polynomial equations

$$\sum_{s,t,u \in B} r_{u,s}^{i,j} r_{t,n}^{s,k} r_{l,m}^{u,t} = \sum_{s,t,u \in B} r_{s,u}^{j,k} r_{l,t}^{i,s} r_{m,n}^{t,u}, \quad i, j, k, l, m, n \in B, \tag{5.133}$$

which can also be represented in a graphical form by assigning elements of the basis B to edges in (5.131) and summing over the elements assigned to the internal edges

$$\sum_{s,t,u\in B} \quad = \quad \sum_{s,t,u\in B} \tag{5.134}$$

with the identifications

$$r^{a,b}_{c,d} = \qquad \forall a, b, c, d \in B. \tag{5.135}$$

Proposition 5.7 *Let* $V = (V, \delta \colon V \to V \otimes C)$ *be a right comodule over a coalgebra* C, *and* $\rho \in (C^{\otimes 2})^*$ *an r-matrix over the coalgebra* C. *Then, the element*

$$\boxed{r} = r := (\mathrm{id}_{V\otimes V} \otimes \rho)(((\mathrm{id}_V \otimes \delta)\sigma_{V,V}) \otimes \mathrm{id}_C)(\mathrm{id}_V \otimes \delta) = \tag{5.136}$$

is an r-matrix over the vector space V. *Here, in the graphical description, the thick lines correspond to* V *and thin lines to* C.

Proof The inverse r^{-1} of r is given by the formula (exercise)

$$r^{-1} = \boxed{r^{-1}} = \tag{5.137}$$

where $\bar{\rho}$ is the convolution inverse of ρ.

By using three times the equality $(\delta \otimes \mathrm{id}_C)\delta = (\mathrm{id}_V \otimes \Delta)$, we transform the left hand side of (5.130) as follows:

$$r_{1,2}r_{2,3}r_{1,2} =$$

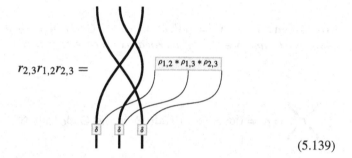

$$(5.138)$$

and, doing a similar calculation for the right hand side of (5.130), we obtain

$$r_{2,3}r_{1,2}r_{2,3} =$$

$$(5.139)$$

thus concluding that Eq. (5.130) is satisfied due to the convolutional Yang–Baxter equality (5.129) for the r-matrix ρ over the coalgebra C. □

The following proposition allows one to view any finite dimensional module over an algebra as a comodule over the restricted dual of that algebra. In this way, one can associate to any finite dimensional representation of a quantum double an r-matrix over the vector space underlying that representation.

Proposition 5.8 *Let V be a finite dimensional left module over an algebra A, and $B \subset V$ a linear basis. Then, V is a right comodule over the coalgebra A^o with the coaction*

$$\delta b = \sum_{b' \in B} b' \otimes \lambda_{b',b} \tag{5.140}$$

where $\{\lambda_{b',b} \mid b, b' \in B\} \subset A^o$ are matrix coefficients of the representation morphism $\lambda: A \to \mathrm{End}(V)$ with respect to the basis B,

$$(\lambda x)b = \sum_{b' \in B} b'\langle\lambda_{b',b}, x\rangle, \quad x \in A, \quad b \in B. \tag{5.141}$$

Proof

(1) We start by checking the equality $(\delta \otimes \mathrm{id}_{A^o})\delta = (\mathrm{id}_V \otimes \Delta)$. Indeed, for any $b \in B$, we have

$$(\delta \otimes \mathrm{id}_{A^o})\delta b = \sum_{b' \in B} (\delta b') \otimes \lambda_{b',b} = \sum_{b' \in B} \left(\sum_{b'' \in B} b'' \otimes \lambda_{b'',b'} \right) \otimes \lambda_{b',b}$$

$$= \sum_{b'' \in B} b'' \otimes \left(\sum_{b' \in B} \lambda_{b'',b'} \otimes \lambda_{b',b} \right) = \sum_{b'' \in B} b'' \otimes (\Delta\lambda_{b'',b})$$

$$= (\mathrm{id}_V \otimes \Delta) \sum_{b'' \in B} b'' \otimes \lambda_{b'',b} = (\mathrm{id}_V \otimes \Delta)\delta b. \qquad (5.142)$$

(2) It remains to check the property $(\mathrm{id}_V \otimes \epsilon)\delta = \mathrm{id}_V$. For any $b \in B$, we calculate

$$(\mathrm{id}_V \otimes \epsilon)\delta b = \sum_{b' \in B} b' \langle \epsilon, \lambda_{b',b} \rangle = \sum_{b' \in B} b' \delta_{b',b} = b. \qquad (5.143)$$

$\square$

Summarizing the contents of Proposition 5.7 and Proposition 5.8, we have the following procedure of constructing a solution of the non-linear system (5.133) of polynomial Yang–Baxter equations.

Let A be an algebra, ρ an r-matrix over the coalgebra A^o (see Definition 5.4), $\lambda: A \to \mathrm{End}(V)$ a finite-dimensional representation, and $B \subset V$ a linear basis. Then, the element $r \in \mathrm{End}(V^{\otimes 2})$ defined by (5.136), which we can also write as

$$r = (\mathrm{id}_{V \otimes V} \otimes \rho)(\mathrm{id}_V \otimes \delta \otimes \mathrm{id}_{A^o})(\sigma_{V,V} \otimes \mathrm{id}_{A^o})(\mathrm{id}_V \otimes \delta), \qquad (5.144)$$

is an r-matrix over the vector space V, where $\delta: V \to V \otimes A^o$ is defined by (5.140) by using the matrix coefficients $\{\lambda_{a,b} \mid a, b \in B\}$ of the representation λ with respect to the basis B (see Eq. (5.141)).

Let us calculate the matrix coefficients $r_{a,b}^{c,d}$ of r (defined in (5.132)) in terms of the evaluation coefficients of ρ.

For any $a, b \in B$, we have

$$r(a \otimes b) = (\mathrm{id}_{V \otimes V} \otimes \rho)(\mathrm{id}_V \otimes \delta \otimes \mathrm{id}_{A^o})(\sigma_{V,V} \otimes \mathrm{id}_{A^o})(\mathrm{id}_V \otimes \delta)(a \otimes b)$$

$$= (\mathrm{id}_{V \otimes V} \otimes \rho)(\mathrm{id}_V \otimes \delta \otimes \mathrm{id}_{A^o})(\sigma_{V,V} \otimes \mathrm{id}_{A^o})(a \otimes \sum_{c \in B} c \otimes \lambda_{c,b})$$

$$= \sum_{c \in B} (\mathrm{id}_{V \otimes V} \otimes \rho)(\mathrm{id}_V \otimes \delta \otimes \mathrm{id}_{A^o})(c \otimes a \otimes \lambda_{c,b})$$

$$= \sum_{c \in B} (\mathrm{id}_{V \otimes V} \otimes \rho)(c \otimes \sum_{d \in B} d \otimes \lambda_{d,a} \otimes \lambda_{c,b}) = \sum_{c,d \in B} c \otimes d \langle \rho, \lambda_{d,a} \otimes \lambda_{c,b} \rangle$$

$$(5.145)$$

so that

$$r_{a,b}^{c,d} = \langle \rho, \lambda_{d,a} \otimes \lambda_{c,b} \rangle \quad \forall a, b, c, d \in B. \tag{5.146}$$

Theorem 5.3 *Let* $\lambda: D(B_q) \rightarrow \mathrm{End}(V)$ *be an irreducible* N*-dimensional representation and* $\{v_n\}_{n \in \underline{N}} \subset V$ *its distinguished linear basis (see Theorem 5.2). Let* $\{\lambda_{m,n}\}_{m,n \in \underline{N}} \subset D(B_q)^o$ *be the matrix coefficients with respect to the basis* $\{v_n\}_{n \in \underline{N}}$ *defined by*

$$(\lambda x) v_n = \sum_{m \in \underline{N}} v_m \langle \lambda_{m,n}, x \rangle, \quad \forall x \in D(B_q). \tag{5.147}$$

Then, the matrix coefficients of the corresponding r-matrix over V are given by

$$r_{l,n}^{m,k} = \langle \lambda_{k,l}, J \iota^o \lambda_{m,n} \rangle$$

$$= \frac{(q^{-1})_n (q)_{N-1-l}}{(q^{-1})_m (q)_{N-1-k} (q)_{n-m}} q^{(n+1-N)k} \xi_q^{N-1-n+k} \xi_{\xi_q}^{-1} \delta_{k+m,l+n} \tag{5.148}$$

if $m \leq n$ *and zero otherwise, see* (4.116) *for the notation.*

Remark 5.4 In what follows, for any generating element $x \in B_q$ (respectively $x \in B_q^o$), we will distinguish it from its image ιx (respectively $J x$) in $D(B_q)$ by putting a dot above it. For example, we will write $\dot{a} \in B_q$ and $a = \iota \dot{a} \in D(B_q)$, $\dot{\psi} \in B_q^o$ and $\psi = J \dot{\psi} \in D(B_q)$, etc. The fact that J reverses the product implies that we have, for example, $\phi \psi = (J \dot{\phi}) J \dot{\psi} = J (\dot{\psi} \dot{\phi})$.

As an intermediate step towards the proof of Theorem 5.3, we first calculate the elements $\iota^o \lambda_{m,n} \in B_q^o$.

Lemma 5.1 *The images* $\iota^o \lambda_{m,n}$, $0 \leq m, n < N$, *as elements of the algebra* B_q^o, *are given by the formula*

$$\iota^o \lambda_{m,n} = \begin{cases} \frac{(q^{-n};q)_{n-m}}{(q)_{n-m}} \dot{\phi}^{n-m} \dot{\theta}_{q^{N-1-n}/\xi_q} & \text{if } m \leq n, \\ 0 & \text{if } m > n \end{cases} \tag{5.149}$$

with the notation defined in (5.113) *and* (4.116).

Proof Recall that for any element $f \in B_q^o$ with the coproduct

$$\Delta f = \sum_{(f)} f_{(1)} \otimes f_{(2)}$$

in Sweedler's sigma notation, we have the decomposition formula (see Eq. (4.143) in the proof of Theorem 4.4)

$$f = \sum_{k \geq 0} \sum_{(f)} \frac{\langle f_{(1)}, \dot{b}^k \rangle}{(q)_k} \dot{\phi}^k (sr)^o f_{(2)} \tag{5.150}$$

which, in the case when $f = \iota^o \lambda_{m,n}$, takes the form

$$\iota^o \lambda_{m,n} = \sum_{k \geq 0} \sum_{l \in \underline{N}} \frac{\langle \iota^o \lambda_{m,l}, \dot{b}^k \rangle}{(q)_k} \dot{\phi}^k (\iota sr)^o \lambda_{l,n}. \tag{5.151}$$

Iteraring the forth formula in (5.86), we obtain

$$\langle \iota^o \lambda_{m,l}, \dot{b}^k \rangle = \langle \lambda_{m,l}, b^k \rangle = (q^{-l}; q)_k \delta_{l,m+k}, \tag{5.152}$$

while iterating the first formula in (5.86) and taking into account the fact that the composed morphism of Hopf algebras $sr : B_q \to B_q$ acts on the basis elements as

$$sr(\dot{b}^i \dot{a}^j) = \delta_{0,i} \dot{a}^j \quad \forall (i, j) \in \omega \times \mathbb{Z}, \tag{5.153}$$

see also (4.119) and (4.120), we obtain

$$\langle (\iota sr)^o \lambda_{l,n}, \dot{b}^i \dot{a}^j \rangle = \langle \lambda_{l,n}, \iota sr(\dot{b}^i \dot{a}^j) \rangle = \delta_{i,0} \langle \lambda_{l,n}, a^j \rangle = \delta_{i,0} \delta_{l,n} \left(q^{N-1-n}/\xi_q \right)^j$$

$$= \delta_{l,n} \langle \dot{\theta}_{q^{N-1-n}/\xi_q}, \dot{b}^i \dot{a}^i \rangle \quad \Rightarrow \quad (\iota sr)^o \lambda_{l,n} = \delta_{l,n} \dot{\theta}_{q^{N-1-n}/\xi_q}. \tag{5.154}$$

Substituting (5.152) and (5.154) into (5.151), we obtain

$$\iota^o \lambda_{m,n} = \sum_{k \geq 0} \sum_{l \in \underline{N}} \frac{(q^{-l}; q)_k \delta_{l,m+k}}{(q)_k} \dot{\phi}^k \delta_{l,n} \dot{\theta}_{q^{N-1-n}/\xi_q}$$

$$= \sum_{k \geq 0} \frac{(q^{-n}; q)_k \delta_{n,m+k}}{(q)_k} \dot{\phi}^k \dot{\theta}_{q^{N-1-n}/\xi_q} = \begin{cases} \frac{(q^{-n};q)_{n-m}}{(q)_{n-m}} \dot{\phi}^{n-m} \dot{\theta}_{q^{N-1-n}/\xi_q} & \text{if } m \leq n, \\ 0 & \text{if } m > n. \end{cases} \tag{5.155}$$

$\square$

Proof of Theorem 5.3 From Lemma 5.1, we obtain

$$J\iota^o \lambda_{m,n} = \frac{(q^{-n}; q)_{n-m}}{(q; q)_{n-m}} \theta_{q^{N-1-n}/\xi_q} \phi^{n-m} \tag{5.156}$$

if $m \leq n$ and zero otherwise. We conclude that $r_{l,n}^{m,k} = 0$ unless $m \leq n$.

In order to handle the case $m \leq n$, we will use the formula

$$\langle \lambda_{k,l}, \phi^m \rangle = (q^{N-k}; q)_m \delta_{k,l+m} \quad \forall k, l, m \in \omega \tag{5.157}$$

which can be obtained by iterating the last formula of (5.86).

Assuming that $m \leq n$, we calculate

$$r^{m,k}_{l,n} = \langle \lambda_{k,l}, J\iota^o \lambda_{m,n} \rangle = \frac{(q^{-n}; q)_{n-m}}{(q; q)_{n-m}} \langle \lambda_{k,l}, \theta_{q^{N-1-n}/\xi_q} \phi^{n-m} \rangle$$

$$= \frac{(q^{-n}; q)_{n-m}}{(q; q)_{n-m}} \left(q^{N-1-n}/\xi_q \right)^{-k} \xi_{q^{N-1-n}/\xi_q} \langle \lambda_{k,l}, \phi^{n-m} \rangle$$

$$= \frac{(q^{-n}; q)_{n-m}(q^{N-k}; q)_{n-m}}{(q)_{n-m}} \left(q^{N-1-n}/\xi_q \right)^{-k} \xi_{q^{N-1-n}/\xi_q} \delta_{k+m,l+n}$$

$$= \frac{(q^{-1})_n (q)_{N-1-l}}{(q^{-1})_m (q)_{N-1-k}(q)_{n-m}} q^{(n+1-N)k} \xi_q^{N-1-n+k} \xi_{\xi_q}^{-1} \delta_{k+m,l+n} \tag{5.158}$$

where, in the third equality, we used an iteration of the third formula in (5.86), in the forth equality, we used (5.157) and, in the last equality, we used the multiplicative property $\xi_u \xi_v = \xi_{uv}$ for any $u, v \in \mathbb{C}_{\neq 0}$, and the identities

$$(q^{-n}; q)_{n-m} = \frac{(q^{-1})_n}{(q^{-1})_m}, \quad 0 \leq m \leq n, \tag{5.159}$$

and

$$(q^{N-k}; q)_{n-m} \Big|_{k+m=l+n} = (q^{N-k}; q)_{k-l} = \frac{(q)_{N-1-l}}{(q)_{N-1-k}}, \quad 0 \leq l \leq k \leq N-1. \tag{5.160}$$

$\square$

Chapter 6
Applications in Knot Theory

In this chapter, we present certain aspects of quantum invariants of knots and links. Mathematically, a *knot* can be defined as a smooth submanifold of the space $\mathbb{R}^3$ (or its compactification $S^3 := \{x \in \mathbb{R}^4 \mid \|x\| = 1\} \simeq \mathbb{R}^3 \cup \{\infty\}$) diffeomorphic to the circle $S^1 := \{x \in \mathbb{R}^2 \mid \|x\| = 1\}$. A *link* is a direct generalisation of a knot when one considers a smooth submanifold of $\mathbb{R}^3$ diffeomorphic to a disjoint union of a finite number of circles. Here are few reasons for general interest in mathematical study of knots and links.

- They allow to visualise intrinsic properties of the space $\mathbb{R}^3$. Therefore, they are an important part of the theory of three dimensional manifolds.
- They are well suited for learning and testing various methods of algebraic and geometric topology.
- There are applications of knots and links in natural sciences, especially in physics, chemistry and molecular biology.

The set $\mathcal{L}$ of all links can be endowed with a natural topology and, among the questions concerning this topological space, the most basic one is the question of the topological *classification* of links: what are the arc-connected components of the topological space $\mathcal{L}$? In the context of this question, we say that two links are *(topologically) equivalent* if they belong to one and the same arc-connected component of the space $\mathcal{L}$. The standard approach to the link classification problem is to construct locally constant functions $f \colon \mathcal{L} \to S$ called *invariants* where S is a set often endowed with an algebraic structure (e.g. groups, rings, etc.).

We will restrict ourselves exclusively to the context of *long knots* which can be thought of as specific smooth submanifolds of the space $\mathbb{R}^3$ diffeomorphic to the real line $\mathbb{R}$. More generally, one can consider also the string links as a disjoint union of finitely many long knots, but it is known that the topological classes of string links are in bijection with the classes of ordinary (closed) links only if the number of components is one, i.e. if a string link is a long knot.

© The Author(s), under exclusive license to Springer Nature Switzerland AG 2023
R. Kashaev, *A Course on Hopf Algebras*, Universitext,
https://doi.org/10.1007/978-3-031-26306-4_6

We will describe in detail the construction of invariants of long knots by using *rigid r-matrices* in monoidal categories. The importance of long knots (as opposed to usual closed knots) will be illustrated by considering a general class of group-theoretical r-matrices put into the context of monoidal categories of relations and spans over sets. These r-matrices are indexed by pointed groups that is groups with a distinguished element (different from the identity element).

As we have seen in the previous Chap. 5, Drinfeld's quantum double construction gives rise to a large class of r-matrices which appear to be rigid, and the associated invariants get factorised through universal invariants associated with the underlying Hopf algebras. Such universal invariants were introduced and studied in a number of works [7, 15, 25–27, 32, 36, 43] mostly either in the context of finite dimensional Hopf algebras or for certain classes of topologically completed infinite dimensional Hopf algebras, for example by considering formal power series. Here, we will define the universal invariants purely algebraically and with minimal assumptions on the underlying Hopf algebras. In particular, we will emphasize the case of infinite dimensional Hopf algebras which can be treated rigorously and purely algebraically due to the approach based on the use of the restricted dual of an algebra in conjunction with the quantum double construction.

Compared to previous chapters, some technical details of proofs in Sect. 6.5, especially those related to functional analysis and integration, are put into exercises.

6.1 Polygonal Links and Diagrams

There exist different equivalent approches for development of the theory of knots and links. Here we briefly describe the so called Piecewise Linear (PL) or polygonal approach. More systematic and detailed explanation can be found, for example, in the book [23].

6.1.1 Polygonal Knots and Links

Let $I := \{t \in \mathbb{R} \mid 0 \le t \le 1\}$ be the closed unit interval of the real line $\mathbb{R}$. For any two points u, v of an $\mathbb{R}$-vector space V, we define the (closed) segment $[u, v] \subset V$ by

$$[u, v] := \{(1 - t)u + tv \mid t \in I\}. \tag{6.1}$$

In particular, we have $[0, 1] = I$. An open segment $]u, v[$ is the interior of the closed segment $[u, v]$:

$$]u, v[:= [u, v] \setminus \{u, v\}. \tag{6.2}$$

Similarly, for any three points $u, v, w \in V$, the (closed) triangle $[u, v, w] \subset V$ is defined by

$$[u, v, w] := \{(1 - s - t)u + sv + tw \mid (s, t) \in I^2, \ s + t \le 1\}. \tag{6.3}$$

Definition 6.1 A subset $K \subset \mathbb{R}^3$ is called *polygonal knot* if it is a piecewise linear simple loop, that is if there exists a finite subset of points $\{p_i\}_{i \in \underline{n}} \subset \mathbb{R}^3$, called *vertices* of K, such that

- $K = \cup_{i \in \underline{n}} [p_i, p_{i+1}]$ with $p_n = p_0$;
- the open segments $]p_i, p_{i+1}[, \ i \in \underline{n}$, are pairwise disjoint and they are disjoint from the set of vertices.

Definition 6.2 A *polygonal link* is a disjoint union of a finite number of polygonal knots $L = \sqcup_{i \in \underline{n}} K_i$ where the knots K_i in this union are called *components* of L.

Definition 6.3 Let $K = \cup_{j \in \underline{n}} [p_j, p_{j+1}]$ be a polygonal knot. We say that a polygonal knot K' is obtained from K by a Δ-*move* if there exist $x \in \mathbb{R}^3 \setminus K$ and $k \in \underline{n}$ such that

$$K' = \left(K \setminus [p_k, p_{k+1}]\right) \cup [p_k, x] \cup [x, p_{k+1}]$$

and $[p_k, x, p_{k+1}] \cap K = [p_k, p_{k+1}]$,

Definition 6.4 Two polygonal links L and L' are said to be *related by a Δ-move* if there are components $K \subset L$ and $K' \subset L'$ such that

- $L' \setminus K' = L \setminus K$;
- one of K and K' is obtained from the other by a Δ-move, and the associated triangle does not intersect the other components of L.

Definition 6.5 Two polygonal links L, L' are said to be Δ-*equivalent* if there exists a finite sequence of polygonal links $L = L_0, L_1, \ldots, L_n = L'$ such that, for any $i \in \underline{n}$, the polygonal links L_i, L_{i+1} are related by a Δ-move.

6.1.2 Link Diagrams

Definition 6.6 Let $p \colon \mathbb{R}^3 \to \mathbb{R}^2$ be the projection on the plane of the first two coordinates $(x, y, z) \mapsto (x, y)$. We say that a polygonal link $L \subset \mathbb{R}^3$ is *generic* if for any $x \in \mathbb{R}^2$, we have $n_x := |p^{-1}(x) \cap L| \in \underline{3}$ with the condition that, for any double point x, i.e. $n_x = 2$, the set $p^{-1}(x) \cap L$ does not contain vertices of L.

Definition 6.7 Let L be a generic polygonal link. The *diagram* of L is the image $p(L) \subset \mathbb{R}^2$ with the information over-under on each double point x to indicate the relative position of the two segments of L containing the two points $p^{-1}(x) \cap L$. The double points with information over-under are called *crossings* of the diagram.

The diagrams corresponding to the components of L are called *components* of the diagram associated with L.

Definition 6.8 We say that two diagrams D and D' are related by a *Reidemeister move* of type R_i, $i \in \{-1, 0, 1, 2, 3\}$, if $D' = (D \setminus F) \cup G$ or $D = (D' \setminus F) \cup G$ where

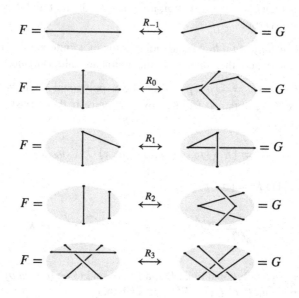

Definition 6.9 Two diagrams D and D' are said to be *R-equivalent* if there exists a finite sequence of diagrams $D = D_0, D_1, \ldots, D_n = D'$ such that, for any $i \in \underline{n}$, the diagrams D_i and D_{i+1} are related by a Reidemeister move.

Exercise 6.1 Show that the following two diagrams are R-equivalent:

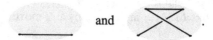

We admit without proof the following Reidemeister theorem (for a proof see, for example, [31]).

Theorem 6.1 (Reidemeister Theorem) *Two generic polygonal links L and L' are Δ-equivalent if and only if the corresponding diagrams are R-equivalent.*

6.1.3 Oriented Links and Diagrams

Definition 6.10 A polygonal link (respectively a diagram) is said to be *oriented* if a direction of travel is chosen for each of its components.

Remark 6.1 The diagram of an oriented generic polygonal link is naturally oriented. The notions of Δ-move, Reidemeister move, Δ-equivalence, R-equivalence, as well as the Reidemeister Theorem, naturally generalise to the context of oriented polygonal links and oriented link diagrams.

Exercise 6.2 Describe all Reidemeister moves of type R_3 for oriented diagrams.

Definition 6.11 Let D be an oriented diagram with the set of crossings C_D. A crossing is said to be *positive* if the ordered pair of vectors $(e_{\sup}, e_{\inf})$ generates the standard orientation of the projection plane. Here $e_{\sup}$ (respectively $e_{\inf}$) corresponds to the oriented upper (respectively lower) strand of the crossing: ⤢. A crossing is said to be *negative* if it is not positive: ⤢.

The *sign map* $\mathrm{sgn}\colon C_D \to \{\pm 1\}$ sends positive crossings to $+1$ and negative crossings to -1.

We denote by $W(D)$ the *writhe* of D defined as the number of positive crossings minus the number of negative crossings:

$$W(D) := \sum_{c \in C_D} \mathrm{sgn}(c) = |\mathrm{sgn}^{-1}(1)| - |\mathrm{sgn}^{-1}(-1)|. \tag{6.4}$$

Remark 6.2 The writhe of a diagram is invariant under all Reidemeister moves with the exception of the oriented versions of the move R_1.

6.2 Long Knots

Subsequently, we will construct knot invariants by using the (polygonal) long knots which correspond to specific piecewise linear embeddings (that is injective maps) of the real line $\mathbb{R}$ into the space $\mathbb{R}^3$, while the usual polygonal knots correspond to piecewise linear embeddings of the circle in $\mathbb{R}^3$. We start by preparing a combinatorial setting for studying long knots.

Let B be the closed unit ball of $\mathbb{R}^3$,

$$B := \{x \in \mathbb{R}^3 \mid \|x\| \le 1\}, \quad \|x\| := \left(\sum x_i^2\right)^{1/2}.$$

Definition 6.12 A subset $K \subset \mathbb{R}^3$ is said to be *long (polygonal) knot* if there exists a finite subset of points $\{p_0, p_1, \ldots, p_n\} \subset B_1$, called *vertices* of K, such that

- $p_0 = (0, -1, 0)$, $p_n = (0, 1, 0)$;
- $K = \big(((\{0\} \times \mathbb{R} \times \{0\}) \setminus B\big) \sqcup \bigcup_{i \in \underline{n}} [p_i, p_{i+1}]$;
- the open segments $]p_i, p_{i+1}[$, $i \in \underline{n}$, are pairwise disjoint and they are disjoint from the set of vertices.

Any long knot has a canonical orientation induced from that of the line $\{0\} \times \mathbb{R} \times \{0\}$:

$$D = \boxed{D}.$$

$$(6.5)$$

All the notions that we have introduced in the previous Sect. 6.1 in the context of oriented polygonal links naturally extend to the context of long knots: Δ-moves, Δ-equivalence, generic long knots and their diagrams, Reidemeister moves, R-equivalence, and the Reidemeister Theorem.

Any long knot K with vertices $\{p_0, \ldots, p_n\}$ corresponds to an oriented polygonal knot $\mathrm{cl}(K)$ called the *closure* of K with the vertices

$$\{p_0, \ldots, p_n, p_{n+1} = (-2, 1, 0), p_{n+2} = (-2, -1, 0)\},$$

and one can show that this correspondence induces a bijection between the respective sets of Δ-equivalence classes. In particular, any invariant of long knots is also an invariant of oriented (closed) polygonal knots.

Two long knot diagrams D and D' can be composed to give another long knot diagram

$$D \circ D' := \boxed{\begin{array}{c} D \\ \hline D' \end{array}}$$

$$(6.6)$$

where the resulting diagram is appropriately (vertically) shifted and rescaled.

In what follows, we assume that the number of vertices of a (long knot) diagram is very large so that the strands of it will be drawn as smooth curves without indication of vertices. For example, if a diagram D represents a long knot K, then the closure $\mathrm{cl}(K)$ is represented by a diagram $\mathrm{cl}(D)$ drawn as follows:

$$D = \boxed{D} \mapsto \mathrm{cl}(D) = \boxed{D}$$

$$(6.7)$$

We will also assume that a long knot diagram is put into a generic position with respect to the vertical axis so that all crossing have non-vertical strands as in the letter X.

Definition 6.13 A (long knot) diagram is called *normal* if it has no local extrema (with respect to the vertical direction) oriented from left to right like ⌢↘ and ⌣↗.

To any diagram D, we associate its *normalization* $\dot{D}$, the diagram obtained from D by the replacements

$$\text{⌢↘} \mapsto \text{⤬} \,, \quad \text{⌣↗} \mapsto \text{⤬} \,.$$

$$(6.8)$$

Example 6.1 The following are examples of normal long knot diagrams for the trefoil and the figure-eight knots:

$$(6.9)$$

It will be of special interest for us the normal long knot diagrams

$$(6.10)$$

and

$$\xi^n := \begin{cases} \uparrow & \text{if } n = 0; \\ \underbrace{\xi^{\mathrm{sgn}(n)} \circ \cdots \circ \xi^{\mathrm{sgn}(n)}}_{|n| \text{ times}} & \text{if } n \neq 0 \end{cases}$$

$$(6.11)$$

where $n \in \mathbb{Z}$, $\mathrm{sgn}(n) := n/|n|$ if $n \neq 0$ and 0 otherwise, and we identify the signs $\pm$ with the numbers ± 1. We have the writhe (see (6.4) for the definition) $W(\xi^n) = 2n$.

Proposition 6.1 *Any normal long knot diagram has an even number of crossings.*

Proof Let D be a normal long knot diagram given in the form of a path composition $\gamma_0 \gamma_1 \cdots \gamma_n \gamma_{n+1}$ where γ_0 and γ_{n+1} are straight infinite half lines, γ_1 the path connecting the point $(0, -1)$ with the first local maximum (counted along the knot), γ_2 the path between the first local maximum and the first local minimum, and so on.

By changing appropriately the types of crossings (positive to negative and vice versa) along γ_1, which do not change the parity of the total number of crossings, we can assume that γ_1 is an overpassing strand so that, by applying the Reidemeister moves of type R_2 and R_3 (which do not change the parity of the total number of crossings) but not the moves R_1, we can pull out the first local maximum and remove all the crossings along γ_1:

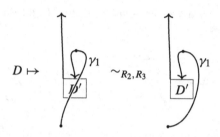

The same reasoning applies to γ_2, namely, by changing the crossings, we can assume that γ_2 is an overpassing strand so that we can pull out the first local minimum and remove all the crossings along γ_2 by applying Reidemeister moves of type R_2 and R_3, and so on. In this way, after n steps, we eventually obtain a normal diagram with no crossings.   □

6.3   Invariants of Long Knots from Rigid r-Matrices

In this section, we introduce the main algebraic input for the construction of long knot invariants, a rigid r-matrix in a monoidal category, and give a detailed description of a long knot invariant associated to a given rigid r-matrix.

Definition 6.14 We say an object G of a monoidal category $\mathcal{C}$ (with tensor product $\otimes$ and unit object I) admits a left *adjoint* if there exists an object F and morphisms

$$\varepsilon: F \otimes G \to I, \qquad \eta: I \to G \otimes F \tag{6.12}$$

such that

$$(\varepsilon \otimes \mathrm{id}_F) \circ (\mathrm{id}_F \otimes \eta) = \mathrm{id}_F, \qquad (\mathrm{id}_G \otimes \varepsilon) \circ (\eta \otimes \mathrm{id}_G) = \mathrm{id}_G. \tag{6.13}$$

In this case, the quadruple $(F, G, \varepsilon, \eta)$ is called *duality* in $\mathcal{C}$.

Example 6.2 In the monoidal category $(\mathbf{Vect}_{\mathbb{F}}, \otimes = \otimes_{\mathbb{F}}, I = \mathbb{F})$ of $\mathbb{F}$-vector spaces, any finite-dimensional vector space V enters the duality $(V^*, V, \mathrm{ev}_V, \mathrm{ev}_V^*)$ where $V^* = \mathrm{L}(V, \mathbb{F})$ is the dual vector space,

$$\mathrm{ev}_V: V^* \otimes V \to \mathbb{F}, \qquad \phi \otimes x \mapsto \phi x, \tag{6.14}$$

is the evaluation map and

$$\mathrm{ev}_V^* : \mathbb{F}^* \simeq \mathbb{F} \to V \otimes V^* \simeq (V^* \otimes V)^* \tag{6.15}$$

the transpose of ev_V also called the coevaluation map. The latter can be given by an explicit formula

$$\mathrm{ev}_V^* \, 1 = \sum_{b \in B} b \otimes b^* \tag{6.16}$$

where $B \subset V$ is a basis and $\{b^*\}_{b \in B} \subset V^*$ is the associated dual basis. □

Definition 6.15 Let $(F, G, \varepsilon, \eta)$ be a duality in a monoidal category $\mathcal{C}$ and a morphism $f \colon A \otimes G \to G \otimes B$ with $A, B \in \mathrm{Ob}\,\mathcal{C}$. The *partial transpose* of f is the morphism $\tilde{f} \colon F \otimes A \to B \otimes F$ defined by

$$\tilde{f} = (\varepsilon \otimes \mathrm{id}_{B \otimes F}) \circ (\mathrm{id}_F \otimes f \otimes \mathrm{id}_F) \circ (\mathrm{id}_{F \otimes A} \otimes \eta).$$

The following definition is Definition 5.5 put into the context of an arbitrary monoidal category.

Definition 6.16 Let $\mathcal{C}$ be a monoidal category. An *r-matrix* over an object $G \in \mathrm{Ob}\,\mathcal{C}$ is an element $r \in \mathrm{Aut}(G \otimes G)$ that satisfies the Yang–Baxter relation

$$r_{1,2}r_{2,3}r_{1,2} = r_{2,3}r_{1,2}r_{2,3}, \quad r_{1,2} := r \otimes \mathrm{id}_G, \quad r_{2,3} := \mathrm{id}_G \otimes r. \tag{6.17}$$

Definition 6.17 Let $(F, G, \varepsilon, \eta)$ be a duality in a monoidal category $\mathcal{C}$. An r-matrix r over G is called *rigid* if the partial transposes

$$\widetilde{r^{\pm 1}} := (\varepsilon \otimes \mathrm{id}_{G \otimes F}) \circ (\mathrm{id}_F \otimes r^{\pm 1} \otimes \mathrm{id}_F) \circ (\mathrm{id}_{F \otimes G} \otimes \eta) \tag{6.18}$$

are invertible.

We remark that the double partial transposes

$$\widetilde{\widetilde{r^{\pm 1}}} := (\varepsilon \otimes \mathrm{id}_{F \otimes F}) \circ (\mathrm{id}_F \otimes \widetilde{r^{\pm 1}} \otimes \mathrm{id}_F) \circ (\mathrm{id}_{F \otimes F} \otimes \eta) \tag{6.19}$$

are invertible with the inverses

$$\left(\widetilde{\widetilde{r^{\pm 1}}}\right)^{-1} = \widetilde{\widetilde{r^{\mp 1}}}. \tag{6.20}$$

Example 6.3 Let H be a finite-dimensional Hopf algebra over a field $\mathbb{F}$ and $h \in \mathrm{Aut}(H)$ a Hopf algebra automorphism. Show that

$$r \colon H \otimes H \to H \otimes H, \quad x \otimes y \mapsto \sum_{(x)} x_{(1)} h((Sx_{(3)})y) \otimes x_{(2)}$$

is a rigid R-matrix in the monoidal category $\mathbf{Vect}_\mathbb{F}$ of vector spaces over a field $\mathbb{F}$ with the tensor product $\otimes = \otimes_\mathbb{F}$ as the monoidal product. $\square$

Associated to a rigid r-matrix r over G with a duality $(F, G, \varepsilon, \eta)$, the *Reshetikhin–Turaev functor* RT_r associates to any normal long knot diagram D the endomorphism $RT_r(D)\colon G \to G$ obtained as follows.

As the non-trivial part of D is contained in $\mathbb{R} \times [-1, 1]$, there exists a finite sequence of real numbers $-1 = t_0 < t_1 < \cdots < t_{n-1} < t_n = 1$ such that, for any $i \in \underline{n}$, the intersection $D_i := D \cap (\mathbb{R} \times [t_i, t_{i+1}])$ is an ordered (from left to right) finite sequence of connected components each of which is isotopic relative to boundary either to one of the four types of segments

$$\uparrow, \ \downarrow, \ \curvearrowleft, \ \smile \tag{6.21}$$

or to one of the eight types of crossings

$$\times, \ \times, \ \times, \ \times, \ \times, \ \times, \ \times, \ \times. \tag{6.22}$$

To such an intersection, we associate a morphism f_i in $\mathcal{C}$ by taking the tensor product (from left to right) of the morphisms associated to the connected fragments of D_i according to the following rules:

$$\uparrow \mapsto \mathrm{id}_G, \quad \downarrow \mapsto \mathrm{id}_F, \quad \curvearrowleft \mapsto \varepsilon, \quad \smile \mapsto \eta, \tag{6.23}$$

$$\times \mapsto r, \quad \times \mapsto r^{-1}, \tag{6.24}$$

$$\times \mapsto \widetilde{r}, \quad \times \mapsto \widetilde{r^{-1}}, \tag{6.25}$$

$$\times \mapsto \widetilde{\widetilde{r}}, \quad \times \mapsto \widetilde{\widetilde{r^{-1}}}, \tag{6.26}$$

$$\times \mapsto \left(\widetilde{r^{-1}}\right)^{-1}, \quad \times \mapsto (\widetilde{r})^{-1}. \tag{6.27}$$

The morphism $RT_r(D)\colon G \to G$ associated to D is obtained as the composition

$$RT_r(D) := f_{n-1} \circ \cdots \circ f_1 \circ f_0. \tag{6.28}$$

Example 6.4 Let $D = \xi^-$ be defined in (6.10). Then $RT_r(D) = f_3 \circ f_2 \circ f_1 \circ f_0$ where

$$f_0 = \mathrm{id}_G \otimes \eta, \quad f_1 = \mathrm{id}_G \otimes (\widetilde{r})^{-1}, \quad f_2 = (\widetilde{r})^{-1} \otimes \mathrm{id}_G, \quad f_3 = \varepsilon \otimes \mathrm{id}_G. \tag{6.29}$$

Theorem 6.2 *Let r be a rigid r-matrix over an object G of a monoidal category* $\mathcal{C}$ *with a duality* $(F, G, \varepsilon, \eta)$. *Then, for any long knot diagram D, the element*

$$J_r(D) := RT_r(\dot{D} \circ \xi^{-w(\dot{D})/2}) \in \text{End}(G) \tag{6.30}$$

depends on only the Reidemeister equivalence class of D.

Proof The proof follows from the functorial properties of the Reshetikhin–Turaev functor, described in [36, 37, 41]. Here we present a proof specifically adapted to the case of normal long knot diagrams.

Let $\sim$ be the equivalence relation on the set of normal long knot diagrams generated by the oriented versions of the Reidemeister moves of types R_2, R_3 and the moves $R_0^{\pm}$ defined by the pictures

$$\tag{6.31}$$

with two possible orientations for the straight segment and two possibilities for crossings. The strategy of the proof is to show first the implication

$$\dot{D} \sim \dot{D}' \Rightarrow RT_r(\dot{D}) = RT_r(\dot{D}') \tag{6.32}$$

which, by taking into account the implication for the writhe

$$\dot{D} \sim \dot{D}' \Rightarrow W(\dot{D}) = W(\dot{D}'), \tag{6.33}$$

ensures the invariance of $J_r(D)$ under all oriented Reidemeister moves R_2 and R_3, and then to verify the invariance under the oriented versions of the Reidemeister move R_1. It is in this last part of the proof where the correction of $\dot{D}$ by $\xi^{-w(\dot{D})/2}$ is crucial.

Invariance of Reshetikhin–Turaev functor with respect to moves $R_0^{\pm}$ follows from the equivalences

$$\tag{6.34}$$

$$\tag{6.35}$$

and the definitions (6.18) and (6.19) of $\widetilde{r^{\pm 1}}$ and $\widetilde{\widetilde{r^{\pm 1}}}$.

Invariance of RT_r with respect to oriented R_2 moves are checked first for eight basic moves

$$\left. \vphantom{X} \right) \sim \left) \left(\sim \right(\overset{RT_r}{\longmapsto} r^{-1} \circ r = \mathrm{id}_{G \otimes G} = r \circ r^{-1}, \tag{6.36}$$

$$\left. \vphantom{X} \right) \sim \left) \left(\sim \right(\overset{RT_r}{\longmapsto} \tilde{r} \circ (\tilde{r})^{-1} = \mathrm{id}_{G \otimes F} = \widetilde{r^{-1}} \circ \left(\widetilde{r^{-1}} \right)^{-1}, \tag{6.37}$$

$$\left. \vphantom{X} \right) \sim \left) \left(\sim \right(\overset{RT_r}{\longmapsto} \left(\widetilde{r^{-1}} \right)^{-1} \circ \widetilde{r^{-1}} = \mathrm{id}_{F \otimes G} = (\tilde{r})^{-1} \circ \tilde{r}, \tag{6.38}$$

$$\left. \vphantom{X} \right) \sim \left) \left(\sim \right(\overset{RT_r}{\longmapsto} \widetilde{\widetilde{r^{-1}}} \circ \widetilde{\tilde{r}} = \mathrm{id}_{F \otimes F} = \widetilde{\tilde{r}} \circ \widetilde{\widetilde{r^{-1}}}, \tag{6.39}$$

and then for two composite moves

$$\bowtie \sim \bowtie \sim \mathcal{O} \sim \mathcal{O} = \subset \tag{6.40}$$

with two possible choices for the crossings.

In order to check invariance of RT_r with respect to the oriented R_3 moves, we introduce a specific parametrization of all those moves. Let Sym(3) be the group of all permutations of the set $\{1, 2, 3\}$. We remark that altogether there are 48 oriented R_3 moves which can be indexed by the set $\mathrm{Sym}(3) \times \{\pm 1\}^3$ as follows.

Given an oriented R_3 move, we enumerate the strands that intervene the move by reading their bottom open ends from left to right

$$\bcancel{} \sim \bcancel{} \quad \rightsquigarrow \quad \underset{1 \quad 2 \quad 3}{\bcancel{}} \sim \underset{1 \quad 2 \quad 3}{\bcancel{}} \tag{6.41}$$

and we define the associated element $(\sigma, \varepsilon) \in \mathrm{Sym}(3) \times \{\pm 1\}^3$ by the conditions that for any $i \in \{1, 2, 3\}$, $\sigma(i)$ is the number of arcs on the i-th strand and $\varepsilon_i = 1$ if i-th strand is oriented upwards. For example, the R_3 move

$$\bcancel{} \sim \bcancel{} \tag{6.42}$$

corresponds to permutation $\sigma = (2, 3) = (1)(2, 3)$ and $\varepsilon = (1, -1, 1)$ while the pair $(\sigma = \mathrm{id}, \varepsilon = (1, 1, 1))$ corresponds to the reference move associated to the

Yang–Baxter relation

$$r_1 \circ r_2 \circ r_1 = r_2 \circ r_1 \circ r_2 \tag{6.43}$$

with the notations $r_1 := r \otimes \mathrm{id}_G$ and $r_2 := \mathrm{id}_G \otimes r$.

We can now show that by using the oriented moves R_2 and $R_0^{\pm}$, any oriented R_3 move is equivalent to the reference move (6.43).

Indeed, in the case $\varepsilon_1 = -1$, we have the equivalences

$$\tag{6.44}$$

which imply the equivalence

$$(\sigma, (-1, \varepsilon_2, \varepsilon_3)) \Leftrightarrow (\sigma \circ (1, 2, 3), (\varepsilon_2, \varepsilon_3, 1)) \tag{6.45}$$

in the set $\mathrm{Sym}(3) \times \{\pm 1\}^3$ thus allowing to reduce the number of negative components of ε. Additionally, a right action of the permutation group $\mathrm{Sym}(3)$ on the set $\mathrm{Sym}(3) \times \{\pm 1\}^3$ is induced by the equivalences

$$\tag{6.46}$$

and

$$\tag{6.47}$$

which correspond to the respective equivalences

$$(\sigma, \varepsilon) \Leftrightarrow (\sigma, \varepsilon) \circ (1, 2) \quad \text{and} \quad (\sigma, \varepsilon) \Leftrightarrow (\sigma, \varepsilon) \circ (2, 3) \tag{6.48}$$

in the set $\mathrm{Sym}(3) \times \{\pm 1\}^3$ where we interpret $(\sigma, \varepsilon) \in \mathrm{Sym}(3) \times \{\pm 1\}^3$ as the map

$$(\sigma, \varepsilon) \colon \{1, 2, 3\} \to \{1, 2, 3\} \times \{\pm 1\}, \quad i \mapsto (\sigma(i), \varepsilon_i). \tag{6.49}$$

Thus, in conjunction with the equivalence (6.45), the right action of the group $\mathrm{Sym}(3)$ on the set $\mathrm{Sym}(3) \times \{\pm 1\}^3$ establishes the equivalence of any oriented R_3

move to the reference move (6.43) and thereby the invariance of RT_r with respect to all oriented R_3 moves.

Finally, in order to prove invariance of J_r with respect to all oriented R_1 moves, we need to check only the invariance with respect to four basic moves of the form

$$\text{✗} \sim \text{∧} \tag{6.50}$$

as all others are consequences of the basic ones and the intermediate equivalence relation $\stackrel{\cdot}{\sim}$ generated by the moves $R_0^{\pm}$, R_2 and R_3:

$$| = \text{∩} \sim \text{✗} \stackrel{\cdot}{\sim} \text{∝} \stackrel{\cdot}{\sim} \text{✗} \Rightarrow \text{∪} \sim \text{✗} \tag{6.51}$$

and

$$| = \text{∩} \sim \text{✗} \stackrel{\cdot}{\sim} \text{✗}. \tag{6.52}$$

Let us analyse the four cases of (6.50) separately.

Case 1 If diagrams D and D' differ by the fragments

$$D \ni \text{✗} , \quad \text{∧} \in D', \tag{6.53}$$

then, by the definition of the normalisation of a long knot diagram, we have the equality $\dot{D} = \dot{D}'$. Thus, $J_r(D) = J_r(D')$.

Case 2 Diagrams D and D' differ by the fragments

$$D \ni \text{✗} , \quad \text{∧} \in D' \tag{6.54}$$

so that the normalised diagrams $\dot{D}$ and $\dot{D}'$ differ by the fragments

$$\dot{D} \ni \text{✗} , \quad \text{✗} \in \dot{D}' \tag{6.55}$$

which imply that

$$W(\dot{D}) = 2 + W(\dot{D}') \Rightarrow \xi^+ \circ \xi^{-W(\dot{D})/2} = \xi^{-W(\dot{D}')/2}. \tag{6.56}$$

On the other hand, we have the equivalence

$$\dot{D} \ni \text{✗} \stackrel{\cdot}{\sim} \text{✗✗} \in \dot{D}' \circ \xi^+ \tag{6.57}$$

which, together with (6.56), implies that

$$\dot{D} \circ \xi^{-W(\dot{D})/2} \sim \dot{D}' \circ \xi^{+} \circ \xi^{-W(\dot{D})/2} = \dot{D}' \circ \xi^{-W(\dot{D}')/2} \Rightarrow J_r(D) = J_r(D').$$
(6.58)

Case 3 Diagrams D and D' differ by the fragments

$$D \ni \text{⚬}, \quad \text{⚬} \in D'$$
(6.59)

so that

$$\dot{D} \ni \text{⚬} \sim \text{⚬} \in \dot{D}' \Rightarrow J_r(D) = J_r(D').$$
(6.60)

Case 4 Diagrams D and D' differ by the fragments

$$D \ni \text{⚬}, \quad \text{⚬} \in D'$$
(6.61)

so that we have for the corresponding normalised diagrams

$$\dot{D} \ni \text{⚬} \sim \text{⚬} \in \dot{D}' \circ \xi^{-}$$
(6.62)

and

$$W(\dot{D}) = W(\dot{D}') - 2 \Rightarrow \xi^{-} \circ \xi^{-W(\dot{D})/2} = \xi^{-W(\dot{D}')/2}.$$
(6.63)

Thus,

$$\dot{D} \circ \xi^{-W(\dot{D})/2} \sim \dot{D}' \circ \xi^{-} \circ \xi^{-W(\dot{D})/2} = \dot{D}' \circ \xi^{-W(\dot{D}')/2} \Rightarrow J_r(D) = J_r(D').$$
(6.64)

□

Exercise 6.3 Let r be a rigid r-matrix over a (finite dimensional) $\mathbb{F}$-vector space V. For any long knot diagram D and any non-zero scalar $\lambda \in \mathbb{F}_{\neq 0}$, show the equality

$$J_r(D) = J_{\lambda r}(D).$$
(6.65)

6.4   Rigid r-Matrices from Racks

In this section, we consider a special class of rigid r-matrices in the categories of relations and spans over sets. Each such r-matrix is associated to a pointed group with a canonical structure of a rack, and Theorem 6.3 identifies the associated invariant with the set of representations of the knot group into the group that underlies the rack.

6.4.1   Categories of Spans and Relations

A *binary relation* from a set X to a set Y as a subset of the cartesian product $X \times Y$. The composition of two binary relations $R \subset X \times Y$ and $S \subset Y \times Z$ is the binary relation $S \circ R \subset X \times Z$ defined by

$$S \circ R := \{(x, z) \in X \times Z \mid \exists y \in Y \colon (x, y) \in R, \ (y, z) \in S\}. \tag{6.66}$$

A *span* from a set X to a set Y is a triple $U = (U, s_U, t_U)$ where U is a set, $s_U \colon U \to X$ and $t_U \colon U \to Y$ are set theoretical maps. The composition of two spans U from X to Y and V from Y to Z is the span from X to Z defined as the pullback space (fibered product) $V \circ U := U \times_Y V$ together with the natural projections to X and Z.

Two spans U and V from X to Y are called *equivalent*, $U \simeq V$, if there exists a bijection $f \colon U \to V$ such that $s_V \circ f = s_U$ and $t_V \circ f = t_U$ so that

$$V = (V, s_V, t_V) \simeq (f^{-1}(V), s_V \circ f, t_V \circ f) = (U, s_U, t_U) = U. \tag{6.67}$$

The composition of spans induces an associative binary operation for the equivalence classes of spans.

Any binary relation $R \subset X \times Y$ is a special case of a span with $s_R \colon R \to X$ and $t_R \colon R \to Y$ being the canonical projections.

Let **Set** be the monoidal category of sets with the cartesian product as the monoidal product, and **Rel** (respectively **Span**) the extension of **Set** with morphisms given by binary relations (respectively equivalence classes of spans). For a morphism $Z \colon X \to Y$ in **Span**, and any $(x, y) \in X \times Y$, we denote

$$Z(x, y) := s_Z^{-1}(x) \cap t_Z^{-1}(y). \tag{6.68}$$

We have a canonical monoidal functor

$$\varpi \colon \mathbf{Span} \to \mathbf{Rel} \tag{6.69}$$

which is identity on the level of objects and for any morphism $Z\colon X \to Y$ in **Span**, the corresponding morphism in **Rel** is given by

$$\varpi(Z) = \{(x, y) \in X \times Y \mid Z(x, y) \neq \emptyset\}. \tag{6.70}$$

Notice that if Z is a relation (as a particular case of spans) then $\varpi(Z) = Z$. In what follows, we concentrate ourselves only in the category **Span**, as the results for **Rel** can always be obtained by applying functor (6.69).

Given a set theoretical map $f\colon X \to Y$, its graph

$$\Gamma_f := \{(x, f(x)) \mid x \in X\} \subset X \times Y, \tag{6.71}$$

being a special case of a binary relation, is naturally interpreted as a morphism $\Gamma_f = (\Gamma_f, p_X, p_Y)\colon X \to Y$ in **Span** where $p_X\colon \Gamma_f \to X$ and $p_Y\colon \Gamma_f \to Y$ are the canonical projections. By using the bijectivity of p_X and the equivalence (6.67) of morphisms in **Span**, we have the equivalence

$$\Gamma_f = (\Gamma_f, p_X, p_Y) \simeq (p_X(\Gamma_f), p_X \circ p_X^{-1}, p_Y \circ p_X^{-1}) = (X, \mathrm{id}_X, f). \tag{6.72}$$

The advantage of the category **Span** (as well as **Rel**) over **Set** is its rigidity, namely, for any set X, the diagonal $\Delta_X := \Gamma_{\mathrm{id}_X}$, interpreted as a morphism in **Span** in two ways

$$\varepsilon_X\colon X \times X \to \{0\} \tag{6.73}$$

and

$$\eta_X\colon \{0\} \to X \times X, \tag{6.74}$$

gives rise to a canonical duality $(X, X, \varepsilon_X, \eta_X)$ in **Span**.

6.4.2 Racks and Rigid r-Matrices in the Category of Spans

The notion of a rack has a rich history, see, for example, [9] and references therein.

Definition 6.18 A *rack* is a set X with a map

$$X^2 \to X^2, \quad (x, y) \mapsto (x \cdot y, y * x), \tag{6.75}$$

such that the binary operation $x \cdot y$ is left self-distributive

$$x \cdot (y \cdot z) = (x \cdot y) \cdot (x \cdot z), \quad \forall(x, y, z) \in X^3, \tag{6.76}$$

and

$$(x \cdot y) * x = y = x \cdot (y * x), \quad \forall (x, y) \in X^2. \tag{6.77}$$

Remark 6.3 A rack satisfying the additional condition $x \cdot x = x$ is known as a *quandle*.

Proposition 6.2 *For any rack* $(X, (x, y) \mapsto (x \cdot y, y * x))$, *the set-theoretical map*

$$r: X^2 \to X^2, \quad (x, y) \mapsto (x \cdot y, x),$$

corresponds to a rigid r-matrix in the category **Span**.

Proof Let us first verify the set-theoretical Yang–Baxter equation. For any $x, y, z \in X$, we have

$$r_{1,2} \circ r_{2,3} \circ r_{1,2}(x, y, z) = r_{1,2} \circ r_{2,3}(x \cdot y, x, z) = r_{1,2}(x \cdot y, x \cdot z, x)$$
$$= ((x \cdot y) \cdot (x \cdot z), x \cdot y, x) \tag{6.78}$$

and

$$r_{2,3} \circ r_{1,2} \circ r_{2,3}(x, y, z) = r_{2,3} \circ r_{1,2}(x, y \cdot z, y) = r_{2,3}(x \cdot (y \cdot z), x, y)$$
$$= (x \cdot (y \cdot z), x \cdot y, x). \tag{6.79}$$

The two results are the same due to the self-distributivity property (6.76).
   The invertibility of r is established by the equality

$$r^{-1} = \bar{r}: X^2 \to X^2, \quad \bar{r}(x, y) = (y, x * y). \tag{6.80}$$

Indeed, for any $x, y \in X$, we have

$$\bar{r} \circ r(x, y) = \bar{r}(x \cdot y, x) = (x, (x \cdot y) * x) = (x, y) \tag{6.81}$$

where the last equality is due to the first equality of (6.77), and

$$r \circ \bar{r}(x, y) = r(y, x * y) = (y \cdot (x * y), y) = (x, y) \tag{6.82}$$

where the last equality is due to the second equality of (6.77).
   The rigidity of the morphism Γ_r in **Span**, see (6.71), follows from the equivalences

$$\widetilde{\Gamma}_r \simeq \Gamma_{r^{-1}}, \quad \widetilde{\Gamma}_{r^{-1}} \simeq \Gamma_{r_{2,1}^{-1}} \tag{6.83}$$

which relate the partial transposes of $\Gamma_{r\pm1}$ in **Span** to graphs of invertible set-theoretical maps. Here, for any set-theoretical map

$$f: X^2 \to X^2, \quad f(x, y) = (f_1(x, y), f_2(x, y)),$$

the map $f_{2,1}: X^2 \to X^2$ is defined by

$$f_{2,1}(x, y) = (f_2(y, x), f_1(y, x)).$$

Exercise 6.4 Show that, for any set-theoretical map $f: Y \times X \to X \times Z$, the partial transpose of Γ_f in **Span**, up to equivalence, is given by the triple

$$\tilde{\Gamma}_f \simeq (Y \times X, (y, x) \mapsto (f_1(y, x), y), (y, x) \mapsto (f_2(y, x), x)) \tag{6.84}$$

with the notation $f(y, x) = (f_1(y, x), f_2(y, x))$.

By using Exercise 6.4, we prove the first equivalence of (6.83) as follows

$$\tilde{\Gamma}_r \simeq (X^2, (x, y) \mapsto (x \cdot y, x), (x, y) \mapsto (x, y))$$

$$= (X^2, r, \mathrm{id}_{X^2}) \simeq (r(X^2), r \circ r^{-1}, r^{-1}) = (X^2, \mathrm{id}_{X^2}, r^{-1}) \simeq \Gamma_{r^{-1}} \tag{6.85}$$

where, in the last equivalence, we used (6.72).

For the second equivalence of (6.83), again by using Exercise 6.4, we have

$$\tilde{\Gamma}_{r^{-1}} \simeq (X^2, (x, y) \mapsto (y, x), (x, y) \mapsto (x * y, y))$$

$$\simeq (X^2, \mathrm{id}_{X^2}, (y, x) \mapsto (x * y, y)) = (X^2, \mathrm{id}_{X^2}, r_{2,1}^{-1}) \simeq \Gamma_{r_{2,1}^{-1}}. \tag{6.86}$$

□

Exercise 6.5 Show that for any rack $(X, (x, y) \mapsto (x \cdot y, y * x))$ the binary operation $x * y$ is right self-distributive

$$(x * y) * z = (x * z) * (y * z). \tag{6.87}$$

6.4.3   Racks Associated to Pointed Groups

Let (G, τ) be a *pointed group* that is a group G together with a fixed element $\tau \in G$. Then, it is verified that the set G with the map

$$G \times G \to G \times G, \quad (g, h) \mapsto (g\tau g^{-1}h, g\tau^{-1}g^{-1}h) \tag{6.88}$$

is a rack, and, thus, it gives rise to a set-theoretical r-matrix

$$r_{G,\tau} \colon G \times G \to G \times G, \quad (g, h) \mapsto (g\tau g^{-1}h, g), \tag{6.89}$$

which, by Proposition 6.2, corresponds to a rigid r-matrix in category **Span**. In this way, through Theorem 6.2, we obtain a long knot invariant $J_{r_{G,\tau}}(D)$. The next theorem reveals the topological content of this invariant.

Theorem 6.3 *There exists a canonical choice of a meridian-longitude pair (m, ℓ) of long knots such that the set $(J_{r_{G,\tau}}(D))(1, \lambda)$ is in bijection with the set of group homomorphisms*

$$\{h \colon \pi_1(\mathbb{R}^3 \setminus f(\mathbb{R}), x_0) \to G \mid h(m) = \tau, \ h(\ell) = \lambda\} \tag{6.90}$$

where $f \colon \mathbb{R} \to \mathbb{R}^3$ is a long knot represented by D.

Proof Let $f \colon \mathbb{R} \to \mathbb{R}^3$ be a long knot whose image under the projection

$$p \colon \mathbb{R}^3 \to \mathbb{R}^2, \quad (x, y, z) \mapsto (x, y). \tag{6.91}$$

is the diagram $\tilde{D} := \dot{D} \circ \xi^{-w(\dot{D})/2}$ with linearly ordered (from bottom to top) set of arcs $a_0, a_1, \ldots, a_n$. As a result, the set of crossings acquires a linear order as well $\{c_i \mid 1 \le i \le n\}$ where c_i is the crossing separating the arcs a_{i-1} and a_i and with the over passing arc a_{κ_i} for a uniquely defined map

$$\kappa \colon \{1, \ldots, n\} \to \{0, \ldots, n\}. \tag{6.92}$$

Let $t_0, t_1, \ldots, t_n \in \mathbb{R}$ be a strictly increasing sequence such that $f(t) = (0, t, 0)$ for all $t \notin [t_0, t_n]$, and for each $i \in \{1, \ldots, n-1\}$, $p(f(t_i))$ belongs to arc a_i and is distinct from any crossing. Choose a base point $x_0 = (0, 0, s)$ with sufficiently large $s \in \mathbb{R}_{>0}$, a sufficiently small $\epsilon \in \mathbb{R}_{>0}$, and define the following paths

$$\alpha_0, \beta_i, \gamma_i \colon [0, 1] \to \mathbb{R}^3, \ i \in \{0, \ldots, n\},$$

$$\alpha_0(t) = (-\epsilon \sin(2\pi t), t_0, \epsilon \cos(2\pi t)), \quad \beta_i(t) = (1 - t)x_0 + (f(t_i) + (0, 0, \epsilon))t,$$

$$\gamma_i(t) = f((1 - t)t_i + tt_0) + (0, 0, \epsilon). \tag{6.93}$$

To each arc a_i of $\tilde{D}$, we associate the homotopy class

$$e_i := [\beta_i \cdot \gamma_i \cdot \bar{\beta}_0] \in \pi_1(\mathbb{R}^3 \setminus f(\mathbb{R}), x_0), \tag{6.94}$$

so that $e_0 = 1$, and the Wirtinger generator

$$w_i := [\beta_i \cdot \gamma_i \cdot \alpha_0 \cdot \bar{\gamma}_i \cdot \bar{\beta}_i] \in \pi_1(\mathbb{R}^3 \setminus f(\mathbb{R}), x_0). \tag{6.95}$$

We have the equalities

$$w_i = e_i w_0 e_i^{-1}, \quad \forall i \in \{0, 1, \ldots, n\}, \tag{6.96}$$

$$e_i = w_{\kappa_i}^{\varepsilon_i} e_{i-1}, \quad \forall i \in \{1, \ldots, n\}, \tag{6.97}$$

where $\varepsilon_i \in \{\pm 1\}$ is the sign of the crossing c_i, and

$$e_i = w_{\kappa_i}^{\varepsilon_i} w_{\kappa_{i-1}}^{\varepsilon_{i-1}} \cdots w_{\kappa_1}^{\varepsilon_1}, \quad \forall i \in \{1, \ldots, n\}. \tag{6.98}$$

We define the canonical meridian-longitude pair (m, ℓ) as follows

$$m := w_0, \quad \ell := e_n. \tag{6.99}$$

Taking into account the condition $\sum_{i=1}^n \varepsilon_i = 0$, we see that ℓ has the trivial image in $H_1(\mathbb{R}^3 \setminus f(\mathbb{R}), \mathbb{Z})$.

Let us show that the following finitely presented groups are isomorphic to the knot group $\pi_1(\mathbb{R}^3 \setminus f(\mathbb{R}), x_0)$:

$$E := \langle m, e_0, \ldots, e_n \mid e_0 = 1, \ e_i = e_{\kappa_i} m^{\varepsilon_i} e_{\kappa_i}^{-1} e_{i-1}, \ 1 \le i \le n \rangle \tag{6.100}$$

and

$$W := \langle w_0, \ldots, w_n \mid w_{\kappa_i}^{\varepsilon_i} w_{i-1} = w_i w_{\kappa_i}^{\varepsilon_i}, \ 1 \le i \le n \rangle. \tag{6.101}$$

As W is nothing else but the Wirtinger presentation of $\pi_1(\mathbb{R}^3 \setminus f(\mathbb{R}), x_0)$, it suffices to see the isomorphism $E \simeq W$. To see the latter, we remark that there are two group homomorphisms

$$u: W \to E, \quad w_i \mapsto e_i m e_i^{-1}, \quad i \in \{0, 1, \ldots, n\}, \tag{6.102}$$

and

$$v: E \to W, \quad m \mapsto w_0, \quad e_0 \mapsto 1, \quad e_i \mapsto w_{\kappa_i}^{\varepsilon_i} w_{\kappa_{i-1}}^{\varepsilon_{i-1}} \cdots w_{\kappa_1}^{\varepsilon_1}, \quad \forall i \in \{1, \ldots, n\}. \tag{6.103}$$

Indeed, we have

$$u(w_{\kappa_i})^{\varepsilon_i} u(w_{i-1}) = u(w_i) u(w_{\kappa_i})^{\varepsilon_i} \Leftrightarrow u(w_{\kappa_i})^{\varepsilon_i} e_{i-1} m e_{i-1}^{-1} = e_i m e_i^{-1} u(w_{\kappa_i})^{\varepsilon_i}$$

$$\Leftrightarrow e_i^{-1} u(w_{\kappa_i})^{\varepsilon_i} e_{i-1} m = m e_i^{-1} u(w_{\kappa_i})^{\varepsilon_i} e_{i-1} \Leftarrow e_i^{-1} u(w_{\kappa_i})^{\varepsilon_i} e_{i-1} = 1$$

$$\Leftrightarrow e_i = u(w_{\kappa_i})^{\varepsilon_i} e_{i-1} \Leftrightarrow e_i = e_{\kappa_i} m^{\varepsilon_i} e_{\kappa_i}^{-1} e_{i-1} \tag{6.104}$$

implying that u is a group homomorphism. We also have

$$v(e_i) = v(e_{\kappa_i})v(m)^{\varepsilon_i}v(e_{\kappa_i})^{-1}v(e_{i-1}) \Leftrightarrow v(e_i)v(e_{i-1})^{-1} = v(e_{\kappa_i})w_0^{\varepsilon_i}v(e_{\kappa_i})^{-1}$$

$$\Leftrightarrow w_{\kappa_i} = v(e_{\kappa_i})w_0 v(e_{\kappa_i})^{-1} \Leftarrow \{w_i = v(e_i)w_0 v(e_i)^{-1}\}_{0 \le i \le n} \qquad (6.105)$$

and, for all $i \in \{1, \ldots, n\}$,

$$v(e_i)w_0 = w_{\kappa_i}^{\varepsilon_i} \cdots w_{\kappa_1}^{\varepsilon_1}w_0 = w_{\kappa_i}^{\varepsilon_i} \cdots w_{\kappa_2}^{\varepsilon_2}w_1 w_{\kappa_1}^{\varepsilon_1} = \cdots$$

$$= w_{\kappa_i}^{\varepsilon_i} \cdots w_{\kappa_k}^{\varepsilon_k}w_{k-1}w_{\kappa_{k-1}}^{\varepsilon_{k-1}} \cdots w_{\kappa_1}^{\varepsilon_1} = \cdots = w_i w_{\kappa_i}^{\varepsilon_i} \cdots w_{\kappa_1}^{\varepsilon_1} = w_i v(e_i) \qquad (6.106)$$

implying that v is a group homomorphism as well.

It remains to show that $v \circ u = \mathrm{id}_W$ and $u \circ v = \mathrm{id}_E$. Indeed, for all $i \in \{0, \ldots, n\}$, we have

$$v(u(w_i)) = v(e_i m e_i^{-1}) = v(e_i)v(m)v(e_i)^{-1} = v(e_i)w_0 v(e_i)^{-1} = w_i \qquad (6.107)$$

implying that $v \circ u = \mathrm{id}_W$. We prove that $u(v(e_i)) = e_i$ for all $i \in \{0, \ldots, n\}$ by recursion on i. For $i = 0$, we have

$$u(v(e_0)) = u(1) = 1 = e_0. \qquad (6.108)$$

Assuming that $u(v(e_{k-1})) = e_{k-1}$ for some $k \in \{1, \ldots, n-1\}$, we calculate

$$u(v(e_k)) = u(w_{\kappa_k}^{\varepsilon_k}v(e_{k-1})) = u(w_{\kappa_k})^{\varepsilon_k}u(v(e_{k-1})) = e_{\kappa_k}m^{\varepsilon_k}e_{\kappa_k}^{-1}e_{k-1} = e_k. \qquad (6.109)$$

Now, any element g of the set $(J_{\Gamma_{G,\tau}}(D))(1, \lambda)$ is a map

$$g: \{0, 1, \ldots, n\} \to G \qquad (6.110)$$

such that

$$g_0 = 1, \ g_n = \lambda, \ g_i = g_{\kappa_i}\tau^{\varepsilon_i}g_{\kappa_i}^{-1}g_{i-1}, \quad \forall i \in \{1, \ldots, n\}. \qquad (6.111)$$

That means that g determines a unique group homomorphism

$$h_g: E \to G \qquad (6.112)$$

such that $h_g(m) = \tau$ and $h_g(e_i) = g_i$ for all $i \in \{0, 1, \ldots, n\}$. On the other hand, any group homomorphism $h: E \to G$ such that $h(m) = \tau$ and $h(\ell) = \lambda$ is of the form $h = h_g$ where $g_i = h(e_i)$. Thus, the map $g \mapsto h_g$ is a set-theoretical bijection between $(J_{\Gamma_{G,\tau}}(D))(1, \lambda)$ and the set of group homomorphisms (6.90). $\qquad \square$

Remark 6.4 Theorem 6.3 illustrates the importance of considering long knots as opposed to closed knots. Namely, by closing a long knot, one identifies two open strands and all the associated data. In particular, one has to impose the equality $\lambda = 1$ that corresponds to considering only those representations where the longitude is realized trivially. This means that in the case of closed diagrams one would obtain less powerful invariants.

6.5 The Alexander Polynomial as a Universal Invariant

In this section, by using the restricted dual of an algebra and Drinfeld's quantum double construction, see Chaps. 4 and 5, we first describe a universal invariant associated to any Hopf algebra with invertible antipode and then illustrate the general construction by the example corresponding to the Hopf algebra B_1 of Sect. 5.4 of Chap. 5. In this way, we will be able to interpret the Alexander polynomial of knots as a universal invariant associated to the Hopf algebra B_1, see Theorem 6.4 below.

6.5.1 Universal Knot Invariants from Hopf Algebras

As follows from Proposition 5.7, any finite-dimensional right comodule over the restricted dual of the quantum double of a Hopf algebra with invertible antipode

$$\delta: V \to V \otimes (D(H))^o, \quad v \mapsto \sum_{(v)} v_{(0)} \otimes v_{(1)}, \tag{6.113}$$

where we extend Sweedler's sigma notation to comodules, gives rise to a rigid r-matrix

$$r_V: V \otimes V \to V \otimes V, \quad u \otimes v \mapsto \sum_{(u),(v)} v_{(0)} \otimes u_{(0)} \langle \varrho, u_{(1)} \otimes v_{(1)} \rangle. \tag{6.114}$$

Exercise 6.6 By using dualities in the monoidal category of vector spaces, see Example 6.2, show that the r-matrix (6.114) is rigid.

Rigidity of r_V implies that there exists a *universal invariant* of long knots $Z_H(K)$ taking its values in the convolution algebra $((D(H))^o)^*$ such that

$$J_{r_V}(K)v = \sum_{(v)} v_{(0)} \langle Z_H(K), v_{(1)} \rangle, \quad \forall v \in V. \tag{6.115}$$

Remark 6.5 The algebra homomorphism $D(H) \to ((D(H))^o)^*$ allows to think of the convolution algebra $((D(H))^o)^*$ as a certain algebra completion of the quantum double $D(H)$ reminiscent of the profinite completion of groups.

6.5.2 The Universal Invariant Associted to B_1

Let $D(B_1)$ be the quantum double of the Hopf algebra B_1 described in Sect. 5.4 of Chap. 5 with the central grouplike element a.

Theorem 6.4 *The universal invariant associated to the Hopf algebra B_1 is of the form $Z_{B_1}(K) = (\Delta_K(a))^{-1}$ where $\Delta_K(t)$ is the Alexander polynomial of K normalised so that $\Delta_K(1) = 1$ and $\Delta_K(t) = \Delta_K(1/t)$.*

Remark 6.6 As an element of the convolution algebra $((D(B_1))^o)^*$, the inverse of the Alexander polynomial $(\Delta_K(a))^{-1}$ in Theorem 6.4 should be interpreted in terms of its Taylor series expansion around $a = 1$ and viewed as an element of the algebra of formal power series $\mathbb{C}[[a-1]] \subset ((D(B_1))^o)^*$,

$$(\Delta_K(a))^{-1} = \sum_{n=0}^{\infty} c_n (a-1)^n, \quad c_n := \frac{1}{n!} \partial^n \Delta_K(t)^{-1} / \partial t^n |_{t=1}. \tag{6.116}$$

The rest of this section is devoted to the proof of Theorem 6.4.

6.5.3 Schrödinger's Coherent States

One of the technical tools in the proof of Theorem 6.4 is the theory of standard Schrödinger's coherent states, which we briefly review here (see, for example, [34]).

For any $n \in \mathbb{Z}_{>0}$, let $H^n \subset L^2(\mathbb{C}^n, \mu_n)$ be the complex Hilbert space of square integrable holomorphic functions $f: \mathbb{C}^n \to \mathbb{C}$ with the scalar product

$$\langle f|g \rangle := \int_{\mathbb{C}^n} \overline{f(z)} g(z) \, d\mu_n(z) \tag{6.117}$$

where the measure $d\mu_n(z)$ on $\mathbb{C}^n$ is given by Lebesgue measure $d\lambda_{2n}(z)$ on $\mathbb{C}^n \simeq \mathbb{R}^{2n}$ multiplied by the Gaussian exponential

$$\frac{1}{\pi^n} e^{-\|z\|^2}, \quad \|z\| := \sqrt{\sum_{i=0}^{n-1} |z_i|^2}. \tag{6.118}$$

Exercise 6.7 Show that the monomials

$$e_k(z) := \prod_{i=0}^{n-1} \frac{z_i^{k_i}}{\sqrt{k_i!}}, \quad k \in \mathbb{Z}_{\geq 0}^n, \tag{6.119}$$

form an orthonormal family of vectors in H^n.

The orthonormal family (6.119) forms a Hilbert basis of H^n due to the convergence of Taylor's (multivariable) expansion for holomorphic functions:

$$f(z) = \sum_{k \in \mathbb{Z}_{\geq 0}^n} \prod_{i=0}^{n-1} \frac{z_i^{k_i}}{k_i!} \frac{\partial^{k_i} f(w)}{\partial w_i^{k_i}}\bigg|_{w=0} = \sum_{k \in \mathbb{Z}_{\geq 0}^n} e_k(z) \prod_{i=0}^{n-1} \frac{1}{\sqrt{k_i!}} \frac{\partial^{k_i} f(w)}{\partial w_i^{k_i}}\bigg|_{w=0}. \tag{6.120}$$

Indeed, in the case when $f \in H^n$, multiplying equality (6.120) by $\overline{e_k(z)}$, integrating over z, using the Fubini (or dominant convergence) theorem and the orthogonality of Exercise 6.7, we obtain

$$\langle e_k | f \rangle = \int_{\mathbb{C}^n} \overline{e_k(z)} f(z) \, d\mu_n(z) = \prod_{i=0}^{n-1} \frac{1}{\sqrt{k_i!}} \frac{\partial^{k_i} f(w)}{\partial w_i^{k_i}}\bigg|_{w=0} \quad \forall k \in \mathbb{Z}_{\geq 0}^n \tag{6.121}$$

so that (6.120) takes the form of an orthogonal expansion in the Hilbert space

$$f = \sum_{k \in \mathbb{Z}_{\geq 0}^n} e_k \langle e_k | f \rangle. \tag{6.122}$$

For any $u \in \mathbb{C}^n$, multiplying both sides of (6.121) by $e_k(u)$, summing over all $k \in \mathbb{Z}_{\geq 0}^n$ and, using the Fubini (or dominant convergence) theorem in the left hand side for exchanging the integration and summation, and the Taylor formula (6.120) in the right hand side, we obtain

$$\int_{\mathbb{C}^n} \varphi_u(\bar{z}) f(z) \, d\mu_n(z) = f(u) \quad \forall (f, u) \in H^n \times \mathbb{C}^n \tag{6.123}$$

where the holomorphic function

$$\varphi_u : \mathbb{C}^n \to \mathbb{C}, \quad z \mapsto \sum_{k \in \mathbb{Z}_{\geq 0}^n} e_k(u) e_k(z) = e^{\sum_{i=0}^{n-1} u_i z_i} \tag{6.124}$$

determines an element $\varphi_u \in H^n$ called *(Schrödinger's) coherent state*. By treating elements of $\mathbb{C}^n$ as column vectors we can write

$$\varphi_u(z) = e^{u^\top z}.$$

Let us also adopt the notation $w^* := \bar{w}^\top$ for the Hermitian conjugation, i.e. the transposition combined with complex conjugation. With this notation we have the equalities

$$\|w\|^2 = w^* w, \quad \varphi_u(\bar{z}) = e^{z^* u}. \tag{6.125}$$

The integral formula (6.123) expresses the reproducing property of coherent states

$$\langle \varphi_{\bar{u}} | f \rangle = f(u) \quad \forall (f, u) \in H^n \times \mathbb{C}^n. \tag{6.126}$$

The choice $f = \varphi_v$ in the last formula gives the scalar product between the coherent states

$$\langle \varphi_{\bar{u}} | \varphi_v \rangle = \varphi_v(u) = \varphi_u(v) = e^{u^\top v}. \tag{6.127}$$

In particular, the norm of a coherent state φ_v is determined by the Euclidean norm of v through the formula

$$\|\varphi_v\| = e^{\|v\|^2/2}. \tag{6.128}$$

6.5.4 A Dense Subspace of H^n

Another useful property of the coherent states is that the (dense) vector subspace A^n of H^n generated by products of coherent states and polynomials is stable under the multiplication of elements of A^n as functions so that A^n carries the additional structure of a commutative algebra, and it is in the domain of any linear differential operator with coefficients in A^n.

Example 6.5 When $n = 1$, the Hilbert basis of H^1 given by the monomials

$$\left\{ e_k(z) = \frac{z^k}{\sqrt{k!}} \mid k \in \mathbb{Z}_{\geq 0} \right\} \subset A^1$$

is the eigenvector basis of the 1-dimensional quantum harmonic oscillator with the (self-adjoint) Hamiltonian operator $z\frac{\partial}{\partial z}$.   □

6.5.5   Gaussian Integration Formula

Writing out explicitly the scalar product in (6.127) as an integral, we obtain an integral identity

$$\int_{\mathbb{C}^n} e^{v^\top z + z^* u} \, d\mu_n(z) = e^{v^\top u} \tag{6.129}$$

which is a special case of the general Gaussian integration formula

$$\int_{\mathbb{C}^n} e^{v^* z + z^* u + z^* M z} \, d\mu_n(z) = \frac{e^{v^* W^{-1} u}}{\det(W)}, \qquad W := I_n - M \tag{6.130}$$

where M is an arbitrary complex n-by-n matrix sufficiently close to zero so that the integral is absolutely convergent. Furthermore, the expansion of (6.130) in power series in M with $u = v = 0$ corresponds to the purely combinatorial MacMahon Master theorem [28].

Exercise 6.8 Prove the Gaussian integration formula (6.130).

6.5.6   Representations of $D(B_1)$ in $A^1[[\hbar]]$

Recall that A^1 is the vector subspace of H^1 generated by products of coherent states with polynomials. Refering to Sect. 5.4 of Chap. 5, for any $\lambda \in \mathbb{C}$, the mappings

$$a \mapsto 1 + \hbar, \quad b \mapsto \frac{\partial}{\partial z}, \quad \phi \mapsto \hbar z, \quad \psi \mapsto \lambda - z \frac{\partial}{\partial z} \tag{6.131}$$

and the action

$$\chi_{u,v} f(z) = e^{\hbar u z} f(vz) \tag{6.132}$$

determine an algebra homomorphism

$$\rho_\lambda : D(B_1) \to \mathrm{End}(A^1[[\hbar]]) \tag{6.133}$$

which sends the central element c defined in (5.121) to $\lambda \hbar$.

An important property of the representation ρ_λ is that the image under $\rho_\lambda^{\otimes 2}$ of the formal r-matrix (5.119) is a well defined element of the algebra $\mathrm{End}(A^1)^{\otimes 2}[[\hbar]]$:

$$\rho_\lambda^{\otimes 2}(R) = (1 + \hbar)^{\lambda - z_0 \frac{\partial}{\partial z_0}} e^{\hbar z_0 \frac{\partial}{\partial z_1}} = \sum_{m,n \geq 0} \frac{\hbar^{m+n}}{n!} \binom{\lambda - z_0 \frac{\partial}{\partial z_0}}{m} \left(z_0 \frac{\partial}{\partial z_1} \right)^n. \tag{6.134}$$

In particular, the double sum in (6.134) truncates to a finite sum if the indeterminate $\hbar$ is nilpotent. Thus, despite the fact that the representation ρ_λ is infinite dimensional, the corresponding r-matrix is well suited for calculation of the image under ρ_λ of the universal invariant $Z_{B_1}(K)$. Moreover, as the parameter λ enters only through the overall normalisation factor $(1 + \hbar)^\lambda$ of the r-matrix, the associated invariant is independent of λ, see Exercise 6.3. For that reason, in what follows, we put $\lambda = 0$ and work only with the representation $\rho := \rho_0$.

In order to apply the construction of Sect. 6.3, we define the input r-matrix

$$r := \rho^{\otimes 2}(R)P \qquad (6.135)$$

where $P \in \text{Aut}(A^2)$ is the permutation operator acting by exchanging the arguments. By using (6.134), we obtain the following explicit action of r:

$$rf(z) = rf(z_0, z_1) = (1 + \hbar)^{-z_0 \frac{\partial}{\partial z_0}} f(z_1 + \hbar z_0, z_0)$$

$$= f\left(z_1 + \frac{\hbar}{1 + \hbar} z_0, \frac{1}{1 + \hbar} z_0\right) = f(U^\top z) \qquad (6.136)$$

where

$$U := \begin{pmatrix} \frac{\hbar}{1+\hbar} & \frac{1}{1+\hbar} \\ 1 & 0 \end{pmatrix} = \begin{pmatrix} 1 - t & t \\ 1 & 0 \end{pmatrix}, \quad t := \frac{1}{1 + \hbar}, \qquad (6.137)$$

is the 2-by-2 matrix entering the definition of the (unrestricted) Burau representation of the braid groups [8]. The action of r on the coherent states is realized by the action of the transposed matrix on the space of parameters:

$$r\varphi_v(z) = \varphi_v(U^\top z) = \varphi_{Uv}(z) \qquad (6.138)$$

where the element $z \in \mathbb{C}^2$ is treated as a column vector. In what follows, we use the indeterminate t defined in terms of $\hbar$ through the formula in (6.137).

6.5.7 The Diagrammatic Rules for the Reshetikhin–Turaev Functor

From the formula (6.138), one calculates the integral kernel of r with respect to the coherent states

$$\langle \varphi_w | r | \varphi_v \rangle = \langle \varphi_{w_0, w_1} | r | \varphi_{v_0, v_1} \rangle = e^{w^* U v} \qquad (6.139)$$

which corresponds to the value of the Reshetikhin–Turaev functor associated to positive crossings of all orientations in normal long knot diagrams with edges coloured by complex numbers:

$$\xrightarrow{RT_r} \langle \varphi_{w_0,w_1} | r | \varphi_{v_0,v_1} \rangle \tag{6.140}$$

Likewise, the integral kernel of r^{-1} given by the formula

$$\langle \varphi_w | r^{-1} | \varphi_v \rangle = \langle \varphi_{w_0,w_1} | r^{-1} | \varphi_{v_0,v_1} \rangle = e^{w^* U^{-1} v} \tag{6.141}$$

is associated with negative crossings of all orientations:

$$\xrightarrow{RT_r} \langle \varphi_{w_0,w_1} | r^{-1} | \varphi_{v_0,v_1} \rangle. \tag{6.142}$$

We complete the list of the diagrammatic rules by adding the rules for vertical segments and local extrema

$$\xrightarrow{RT_r} e^{\bar{w} v} \tag{6.143}$$

where $e^{\bar{w} v}$ is the integral kernel of the identity operator id_{A^1}:

$$\langle \varphi_w | \mathrm{id}_{A^1} | \varphi_v \rangle = \langle \varphi_w | \varphi_v \rangle = e^{\bar{w} v}. \tag{6.144}$$

For later use, we calculate the following two Reshetikhin–Turaev images

$$\langle \varphi_w | RT_r \left(\begin{array}{c} \end{array} \right) | \varphi_v \rangle = RT_r \left(\begin{array}{c} \end{array} \right) = \int_{\mathbb{C}} \langle \varphi_{w,u} | r | \varphi_{v,u} \rangle \, \mathrm{d}\mu_1(u)$$

$$= \int_{\mathbb{C}} e^{(\bar{w}\ \bar{u}) \left(\begin{smallmatrix} 1-t & t \\ 1 & 0 \end{smallmatrix} \right) \left(\begin{smallmatrix} v \\ u \end{smallmatrix} \right)} \, \mathrm{d}\mu_1(u) = \int_{\mathbb{C}} e^{\bar{w}(1-t)v + \bar{w}tu + \bar{u}v} \, \mathrm{d}\mu_1(u) = e^{\bar{w} v} \tag{6.145}$$

and

$$\langle \varphi_w | RT_r \left(\vcenter{\hbox{⟲}} \right) | \varphi_v \rangle = RT_r \left(\vcenter{\hbox{⟲}} \right) = \int_{\mathbb{C}} \langle \varphi_{w,u} | r^{-1} | \varphi_{v,u} \rangle \, d\mu_1(u)$$

$$= \int_{\mathbb{C}} e^{(\bar{w}\ \bar{u})\left(\begin{smallmatrix} 0 & 1 \\ t^{-1} & 1-t^{-1} \end{smallmatrix} \right)\left(\begin{smallmatrix} v \\ u \end{smallmatrix} \right)} \, d\mu_1(u) = \int_{\mathbb{C}} e^{\bar{w}u + \bar{u}t^{-1}v + \bar{u}(1-t^{-1})u} \, d\mu_1(u) = t e^{\bar{w}v}$$

$$(6.146)$$

where the integrals are calculated by using the Gaussian integration formula (6.130).

Proof of Theorem 6.4 Let a knot K be represented by the closure of a braid $\beta \in B_n$. Let us choose a normal long knot diagram D_β representing K according to the picture

$$W(D_\beta) = W(\beta) + n - 1 \qquad (6.148)$$

which is an even number by Proposition 6.1. Taking into account the value (6.145), writing the matrix $\psi_n(\beta)$ in the block form

$$\psi_n(\beta) = \begin{pmatrix} \hat{\beta}_n & b_\beta \\ c_\beta & d_\beta \end{pmatrix}, \qquad (6.149)$$

and using the general Gaussian integration formula (6.130), we calculate

$$\langle \varphi_w | RT_r(D_\beta) | \varphi_v \rangle = RT_r \left(\boxed{D_\beta} \right) = \int_{\mathbb{C}^{n-1}} e^{(u^* \ \bar{w}) \psi_n(\beta) \left(\begin{smallmatrix} u \\ v \end{smallmatrix} \right)} \, d\mu_{n-1}(u)$$

$$= \int_{\mathbb{C}^{n-1}} e^{(u^* \ \bar{w}) \left(\begin{smallmatrix} \hat{\beta}_n & b_\beta \\ c_\beta & d_\beta \end{smallmatrix} \right)\left(\begin{smallmatrix} u \\ v \end{smallmatrix} \right)} \, d\mu_{n-1}(u)$$

$$= \int_{\mathbb{C}^{n-1}} e^{\bar{w}d_\beta v + \bar{w}c_\beta u + u^* b_\beta v + u^* \hat{\beta}_n u} \, d\mu_{n-1}(u) = \frac{e^{\bar{w}d_\beta v + \bar{w}c_\beta (I_{n-1} - \hat{\beta}_n)^{-1} b_\beta v}}{\det(I_{n-1} - \hat{\beta}_n)}. \qquad (6.150)$$

On the other hand, given the fact that we are calculating a central element realised by a scalar so that on a priori grounds the result should be proportional to the integral kernel of the identity operator $e^{\tilde{w}v}$, we conclude that the identity

$$d_\beta + c_\beta(I_{n-1} - \hat{\beta}_n)^{-1}b_\beta = 1 \qquad (6.151)$$

is satisfied for the matrix $\psi_n(\beta)$, a property which does not look to be easy to prove without passing through the Gaussian integration and referring to the universal invariant.

Finally, it remains to take into account the writhe correction, which, according to the values in (6.145) and (6.146) is given by the formula

$$\langle \varphi_w | RT_r \left(\xi^{-W(D_\beta)/2} \right) | \varphi_v \rangle e^{-\tilde{w}v} = t^{W(D_\beta)/2} = t^{(W(\beta)+n-1)/2} \qquad (6.152)$$

where we use the notation ξ^k from Sect. 6.3 for a specific class of long knot diagrams used to compensate the writhe of the diagram. Putting together (6.150) and (6.152), the result for the invariant $J_r(K)$ reads

$$\langle \varphi_w | J_r(K) | \varphi_v \rangle e^{-\tilde{w}v} = \frac{t^{(W(\beta)+n-1)/2}}{\det(I_{n-1} - \hat{\beta}_n)} = \frac{1}{\Delta_K(t)} \qquad (6.153)$$

where the last equality is due to formula (6.167) of Theorem 6.5. Taking into account the relation between $\hbar$, t and the realisation of the central element a of $D(B_1)$ as well as the symmetry of the Alexander polynomial under the substitution $t \mapsto t^{-1}$, we conclude the proof.   □

Remark 6.7 Theorem 6.4 is consistent with the Melvin–Morton–Rozansky conjecture, proven by Bar-Natan and Garoufalidis in [4] and by Garoufalidis and Lê in [14], stating that the n-th colored Jones polynomial in the limit $n \to \infty$ with $q = t^{1/n}$ and fixed t, tends to $(\Delta_K(t))^{-1}$. Indeed, for the non-commutative Hopf algebra B_q with q not a root of unity, the quantum double $D(B_q)$ is closely related to the quantum group $U_q(sl_2)$. In particular, for each $n \in \mathbb{Z}_{>0}$, it admits an n-dimensional irreducible representation corresponding to the n-th colored Jones polynomial. In the limit $n \to \infty$ with $q = t^{1/n}$ and fixed t, one recovers an infinite-dimensional representation of the Hopf algebra B_1 where the central element a takes the value t.

6.6 The Alexander Polynomial from the Burau Representation

In this section, we give a proof of Theorem 6.5 below which is used in the proof of Theorem 6.4. Originally, Theorem 6.5 is proven in the work [22]. We adopt the notation of [20] and first briefly describe the unreduced and reduced Burau representations of the braid groups B_n for $n \geq 2$.

For any $k \geq 1$, denote by I_k the identity $k \times k$ matrix. Let

$$\psi_n : B_n \to \mathrm{GL}_n(\Lambda), \quad \Lambda := \mathbb{Z}[t^{\pm 1}], \tag{6.154}$$

be the unrestricted Burau representation where Artin's standard generators σ_i, $1 \leq i < n$, are realised by the matrices

$$\psi_n(\sigma_i) = U_i := I_{i-1} \oplus U \oplus I_{n-i-1}. \tag{6.155}$$

For any $k \geq 1$, define the invertible upper triangular $k \times k$ matrix

$$C_k = \sum_{1 \leq i \leq j \leq k} E_{i,j} = I_k + \sum_{1 \leq i < j \leq k} E_{i,j} \tag{6.156}$$

where $E_{i,j}$ is the matrix with the only non-zero element 1 at the place (i, j). Its inverse has the form

$$C_k^{-1} = I_k - \sum_{i=1}^{n-1} E_{i,i+1}. \tag{6.157}$$

Indeed, one easily calculates

$$C_k \Big(I_k - \sum_{i=1}^{n-1} E_{i,i+1} \Big) = C_k - \sum_{1 \leq i < j \leq k} E_{i,j} = I_k. \tag{6.158}$$

We remark on the block structure of $C_k^{\pm 1}$:

$$C_k = \begin{pmatrix} C_{k-1} & 1_{k-1} \\ 0_{k-1}^\top & 1 \end{pmatrix}, \quad C_k^{-1} = \begin{pmatrix} C_{k-1}^{-1} & -C_{k-1}^{-1} 1_{k-1} \\ 0_{k-1}^\top & 1 \end{pmatrix} = \begin{pmatrix} C_{k-1}^{-1} & 0_{k-2} \\ 0_{k-1}^\top & -1 \\ & 1 \end{pmatrix} \tag{6.159}$$

where 0_i (respectively 1_i) is the column of length i composed of 0's (respectively of 1's) and, in the last equality, we use the relation

$$C_k^{-1} 1_k = \begin{pmatrix} 0_{k-1} \\ 1 \end{pmatrix}. \tag{6.160}$$

As is shown in [20], for any $\beta \in B_n$, one has the equality

$$C_n^{-1} \psi_n(\beta) C_n = \begin{pmatrix} \psi_n^r(\beta) & 0_{n-1} \\ *_\beta & 1 \end{pmatrix} \tag{6.161}$$

where $\psi_n^r : B_n \to \mathrm{GL}_{n-1}(\Lambda)$ is the reduced Burau representation, and $*_\beta$ is a row of length $n-1$ over Λ linearly depending on the rows a_i, $1 \le i \le n-1$, of the matrix $\psi_n^r(\beta) - I_{n-1}$ through the formula[1]

$$(1 - t^n)*_\beta = \sum_{i=1}^{n-1}(t^i - 1)a_i. \tag{6.162}$$

Lemma 6.1 *Let $\hat{\beta}_n$ be the $(n-1) \times (n-1)$ matrix obtained from $\psi_n(\beta)$ by throwing away the n-th row and the n-th column. Then, one has the following equality in Λ:*

$$(t^{-n} - 1)\det(\hat{\beta}_n - I_{n-1}) = (t^{-1} - 1)\det(\psi_n^r(\beta) - I_{n-1}). \tag{6.163}$$

Proof We have the following equality of matrices:

$$\hat{\beta}_n = (C_{n-1}\psi_n^r(\beta) + 1_{n-1}*_\beta)C_{n-1}^{-1}$$

$$\Leftrightarrow \quad C_{n-1}^{-1}\hat{\beta}_n C_{n-1} = \psi_n^r(\beta) + C_{n-1}^{-1}1_{n-1}*_\beta = \psi_n^r(\beta) + \begin{pmatrix} 0_{n-2} \\ 1 \end{pmatrix}*_\beta . \tag{6.164}$$

One can verify this by explicit calculation based on the block structure (6.159):

$$\psi_n(\beta) = \begin{pmatrix} C_{n-1} & 1_{n-1} \\ 0_{n-1}^{\mathsf{T}} & 1 \end{pmatrix} \begin{pmatrix} \psi_n^r(\beta) & 0_{n-1} \\ *_\beta & 1 \end{pmatrix} C_n^{-1}$$

$$= \begin{pmatrix} C_{n-1}\psi_n^r(\beta) + 1_{n-1}*_\beta & * \\ *_\beta & 1 \end{pmatrix} \begin{pmatrix} C_{n-1}^{-1} & * \\ 0_{n-1}^{\mathsf{T}} & 1 \end{pmatrix}$$

$$= \begin{pmatrix} (C_{n-1}\psi_n^r(\beta) + 1_{n-1}*_\beta)C_{n-1}^{-1} & * \\ * & 1 \end{pmatrix}. \tag{6.165}$$

Thus,

$$\det(\hat{\beta}_n - I_{n-1}) = \det\left(\psi_n^r(\beta) - I_{n-1} + \begin{pmatrix} 0_{n-2} \\ 1 \end{pmatrix}*_\beta \right) = \det\begin{pmatrix} a_1 \\ \vdots \\ a_{n-2} \\ a_{n-1}+*_\beta \end{pmatrix}. \tag{6.166}$$

---

[1] This is the content of Lemma 3.10 of [20] where the formula is written with a typo.

By multiplying both sides of (6.166) by $(1 - t^n)$ and using (6.162), we obtain

$$(1-t^n)\det(\hat{\beta}_n - I_{n-1}) = \det\begin{pmatrix} a_1 \\ \vdots \\ a_{n-2} \\ (1-t^n)(a_{n-1}+*_\beta) \end{pmatrix} = \det\begin{pmatrix} a_1 \\ \vdots \\ a_{n-2} \\ (1-t^n)a_{n-1}+\sum_{i=1}^{n-1}(t^i-1)a_i \end{pmatrix}$$

$$= \det\begin{pmatrix} a_1 \\ \vdots \\ a_{n-2} \\ (t^{-1}-1)t^n a_{n-1} \end{pmatrix} = (t^{-1} - 1)t^n \det(\psi_n^r(\beta) - I_{n-1})$$

where, in the third equality, we dropped from the sum all the terms proportional to the rows different from $n - 1$. □

Theorem 6.5 *Let a knot K be the closure of a braid $\beta \in B_n$ and $\psi_n(\beta) \in$ $GL_n(\mathbb{Z}[t^{\pm1}])$ the image of β under the unrestricted Burau representation (where the images of the standard Artin generators are linear in t). Let $\hat{\beta}_n$ be the $(n-1)\times(n-1)$ matrix obtained from $\psi_n(\beta)$ by throwing away the n-th column and the n-th row. Then, the Alexander polynomial of K is given by the formula*

$$\Delta_K(t) = t^{\frac{1-n-W(\beta)}{2}} \det(I_{n-1} - \hat{\beta}_n) \tag{6.167}$$

where I_k denotes the identity $k \times k$ matrix and the writhe $W: B_n \to \mathbb{Z}$ can be seen as the group homomorphism that sends the Artin generators to 1.

Proof We have the following formula for the Alexander polynomial proven in [20, Theorem 3.13]:

$$\Delta_K(t) = (-1)^{n-1}t^{(n-1-W(\beta))/2}\frac{t-1}{t^n-1}\det(\psi_n^r(\beta) - I_{n-1}) \tag{6.168}$$

which is equivalent to (6.167) due to Lemma 6.1. □

References

1. Abe, E.: Hopf Algebras. Cambridge Tracts in Mathematics, vol. 74. Cambridge University Press, Cambridge (1980). Translated from the Japanese by Hisae Kinoshita and Hiroko Tanaka
2. Apostol, T.M.: Introduction to Analytic Number Theory. Undergraduate Texts in Mathematics. Springer, New York (1976)
3. Atiyah, M.: Topological quantum field theories. Inst. Hautes Études Sci. Publ. Math. **68**, 175–186 (1989)
4. Bar-Natan, D., Garoufalidis, S.: On the Melvin-Morton-Rozansky conjecture. Invent. Math. **125**(1), 103–133 (1996)
5. Baxter, R.J.: Partition function of the eight-vertex lattice model. Ann. Phys. **70**, 193–228 (1972)
6. Baxter, R.J.: Exactly Solved Models in Statistical Mechanics. Academic Press, [Harcourt Brace Jovanovich, Publishers], London (1982)
7. Bruguières, A., Virelizier, A.: Hopf diagrams and quantum invariants. Algebr. Geom. Topol. **5**, 1677–1710 (2005)
8. Burau, W.: Über Zopfgruppen und gleichsinnig verdrillte Verkettungen. Abh. Math. Sem. Univ. Hamburg **11**(1), 179–186 (1935)
9. Carter, J.S.: A survey of quandle ideas. In: Introductory Lectures on Knot Theory. Series on Knots and Everything, vol. 46, pp. 22–53. World Science Publishers, Hackensack (2012)
10. Dăscălescu, S., Năstăsescu, C., Raianu, Ş.: Hopf Algebras. Monographs and Textbooks in Pure and Applied Mathematics, vol. 235. Marcel Dekker, New York (2001). An introduction
11. Drinfel'd, V.G.: Quantum groups. In: Proceedings of the International Congress of Mathematicians, (Berkeley, CA, 1986), vol. 1, 2, pp. 798–820. American Mathematical Society, Providence (1987)
12. Faddeev, L.: Integrable models in $(1 + 1)$-dimensional quantum field theory. In: Recent Advances in Field Theory and Statistical Mechanics (Les Houches, 1982), pp. 561–608. North-Holland, Amsterdam (1984)
13. Faddeev, L.D., Reshetikhin, N.Y., Takhtajan, L.A.: Quantization of Lie groups and Lie algebras. In: Algebraic Analysis, vol. I, pp. 129–139. Academic Press, Boston (1988)
14. Garoufalidis, S., Lê, T.T.Q.: Asymptotics of the colored Jones function of a knot. Geom. Topol. **15**(4), 2135–2180 (2011)
15. Habiro, K.: Bottom tangles and universal invariants. Algebr. Geom. Topol. **6**, 1113–1214 (2006)
16. Jantzen, J.C.: Lectures on Quantum Groups. Graduate Studies in Mathematics, vol. 6. American Mathematical Society, Providence (1996)
17. Jimbo, M.: A q-difference analogue of $U(g)$ and the Yang-Baxter equation. Lett. Math. Phys. **10**(1), 63–69 (1985)

© The Author(s), under exclusive license to Springer Nature Switzerland AG 2023
R. Kashaev, *A Course on Hopf Algebras*, Universitext,
https://doi.org/10.1007/978-3-031-26306-4

18. V.F.R. Jones, Hecke algebra representations of braid groups and link polynomials. Ann. Math. **126**(2), 335–388 (1987)
19. Kassel, C.: Quantum Groups. Graduate Texts in Mathematics, vol. 155. Springer, New York (1995)
20. Kassel, C., Turaev, V.: Braid Groups. Graduate Texts in Mathematics, vol. 247. Springer, New York (2008). With the graphical assistance of Olivier Dodane
21. Kassel, C., Rosso, M., Turaev, V.: Quantum Groups and Knot Invariants. Panoramas et Synthèses [Panoramas and Syntheses], vol. 5. Société Mathématique de France, Paris (1997)
22. Kauffman, L.H., Saleur, H.: Free fermions and the Alexander-Conway polynomial. Commun. Math. Phys. **141**(2), 293–327 (1991)
23. Kawauchi, A.: A Survey of Knot Theory. Birkhäuser Verlag, Basel (1996). Translated and revised from the 1990 Japanese original by the author
24. Kuliš, P.P., Rešetihin, N.J.: Quantum linear problem for the sine-Gordon equation and higher representations. Zap. Nauchn. Sem. Leningrad. Otdel. Mat. Inst. Steklov. **101**, 101–110, 207 (1981). Questions in quantum field theory and statistical physics, 2
25. Lawrence, R.J.: A universal link invariant using quantum groups. In: Differential Geometric Methods in Theoretical Physics (Chester, 1988), pp. 55–63. World Scientific Publishing, Teaneck (1989)
26. Lee, H.C.: Tangles, links and twisted quantum groups. In: Physics, Geometry, and Topology (Banff, AB, 1989). NATO Advanced Science Institutes Series B: Physics, vol. 238, pp. 623–655. Plenum, New York (1990)
27. Lyubashenko, V.: Tangles and Hopf algebras in braided categories. J. Pure Appl. Algebra **98**(3), 245–278 (1995)
28. MacMahon, P.A.: Combinatory Analysis. Two volumes (bound as one). Chelsea Publishing, New York (1960)
29. Majid, S.: Foundations of Quantum Group Theory. Cambridge University Press, Cambridge (1995)
30. McGuire, J.B.: Study of exactly soluble one-dimensional N-body problems. J. Math. Phys. **5**, 622–636 (1964)
31. Murasugi, K.: Knot Theory and its Applications. Birkhäuser, Boston (1996). Translated from the 1993 Japanese original by Bohdan Kurpita
32. Ohtsuki, T.: Colored ribbon Hopf algebras and universal invariants of framed links. J. Knot Theory Ramifications **2**(2), 211–232 (1993)
33. Onsager, L.: Crystal statistics. I. A two-dimensional model with an order-disorder transition. Phys. Rev. **65**, 117–149 (1944)
34. Perelomov, A.: Generalized Coherent States and Their Applications. Texts and Monographs in Physics. Springer, Berlin (1986)
35. Radford, D.E.: Hopf Algebras. Series on Knots and Everything, vol. 49. World Scientific Publishing, Hackensack (2012)
36. Reshetikhin, N.Y.: Quasitriangular Hopf algebras and invariants of links. Algebra i Analiz **1**(2), 169–188 (1989)
37. Reshetikhin, N.Y., Turaev, V.G.: Ribbon graphs and their invariants derived from quantum groups. Commun. Math. Phys. **127**(1), 1–26 (1990)
38. Segal, G.B.: The definition of conformal field theory. In: Differential Geometrical Methods in Theoretical Physics (Como, 1987). NATO Advanced Science Institutes Series C: Mathematical and Physical Sciences, vol. 250, pp. 165–171. Kluwer Academic Publishers, Dordrecht (1988)
39. Sweedler, M.E.: Hopf Algebras. Mathematics Lecture Note Series. W. A. Benjamin, New York (1969)
40. Takeuchi, M.: Free Hopf algebras generated by coalgebras. J. Math. Soc. Japan **23**, 561–582 (1971)
41. Turaev, V.G.: Quantum Invariants of Knots and 3-Manifolds. de Gruyter Studies in Mathematics, vol. 18. Walter de Gruyter, Berlin (1994)
42. Turaev, V., Virelizier, A.: Monoidal Categories and Topological Field Theory. Progress in Mathematics, vol. 322. Birkhäuser/Springer, Cham (2017)

43. Virelizier, A.: Kirby elements and quantum invariants. Proc. Lond. Math. Soc. **93**(2), 474–514 (2006)
44. Waterhouse, W.C.: Introduction to Affine Group Schemes. Graduate Texts in Mathematics, vol. 66. Springer, New York (1979)
45. Witten, E.: Topological quantum field theory. Commun. Math. Phys. **117**(3), 353–386 (1988)
46. Yang, C.N.: Some exact results for the many-body problem in one dimension with repulsive delta-function interaction. Phys. Rev. Lett. **19**, 1312–1315 (1967)
47. Zamolodchikov, A.B., Zamolodchikov, A.B.: Factorized S-matrices in two dimensions as the exact solutions of certain relativistic quantum field theory models. Ann. Phys. **120**(2), 253–291 (1979)

Index

Printed in the United States
by Baker & Taylor Publisher Services